T0224044

SpringerBriefs in Applied Sciences and Technology

SpringerBriefs present concise summaries of cutting-edge research and practical applications across a wide spectrum of fields. Featuring compact volumes of 50 to 125 pages, the series covers a range of content from professional to academic.

Typical publications can be:

A timely report of state-of-the art methods

An introduction to or a manual for the application of mathematical or computer techniques

A bridge between new research results, as published in journal articles

A snapshot of a hot or emerging topic

An in-depth case study

A presentation of core concepts that students must understand in order to make independent contributions

SpringerBriefs are characterized by fast, global electronic dissemination, standard publishing contracts, standardized manuscript preparation and formatting guidelines, and expedited production schedules.

On the one hand, SpringerBriefs in Applied Sciences and Technology are devoted to the publication of fundamentals and applications within the different classical engineering disciplines as well as in interdisciplinary fields that recently emerged between these areas. On the other hand, as the boundary separating fundamental research and applied technology is more and more dissolving, this series is particularly open to trans-disciplinary topics between fundamental science and engineering.

Indexed by EI-Compendex and Springerlink

More information about this series at http://www.springer.com/series/8884

Stepan S. Batsanov · Evgeny D. Ruchkin
Inga A. Poroshina

Refractive Indices of Solids

 Springer

Stepan S. Batsanov
National Research Institute
  for Physical-Technical and Radiotechnical
  Measurements
Mendeleyevo
Russia

Evgeny D. Ruchkin
National Research Institute
  for Physical-Technical and Radiotechnical
  Measurements
Mendeleyevo
Russia

Inga A. Poroshina
Novosibirsk State Pedagogical University
Novosibirsk
Russia

ISSN 2191-530X   ISSN 2191-5318   (electronic)
SpringerBriefs in Applied Sciences and Technology
ISBN 978-981-10-0796-5   ISBN 978-981-10-0797-2   (eBook)
DOI 10.1007/978-981-10-0797-2

Library of Congress Control Number: 2016946003

© The Author(s) 2016
This work is subject to copyright. All rights are reserved by the Publisher, whether the whole or part of the material is concerned, specifically the rights of translation, reprinting, reuse of illustrations, recitation, broadcasting, reproduction on microfilms or in any other physical way, and transmission or information storage and retrieval, electronic adaptation, computer software, or by similar or dissimilar methodology now known or hereafter developed.
The use of general descriptive names, registered names, trademarks, service marks, etc. in this publication does not imply, even in the absence of a specific statement, that such names are exempt from the relevant protective laws and regulations and therefore free for general use.
The publisher, the authors and the editors are safe to assume that the advice and information in this book are believed to be true and accurate at the date of publication. Neither the publisher nor the authors or the editors give a warranty, express or implied, with respect to the material contained herein or for any errors or omissions that may have been made.

Printed on acid-free paper

This Springer imprint is published by Springer Nature
The registered company is Springer Science+Business Media Singapore Pte Ltd.

Foreword

Refractometry, i.e. measurements of refractive indices (RIs) (n) of glasses, fine powders and amorphous solids at normal or elevated temperatures and pressures, provides the information that is often in accessible to other physical methods. In particular, refractometry is successfully applied to examining the effects of shock-wave compression on condensed matter, both during the compression itself (which lasts less than a microsecond!) and in the samples recovered after unloading. Most recently, a method of measuring RIs of nanoparticles in colloidal solutions was developed, allowing to determine their composition and structure. Applications of refractometry to the study of electronic structure of simple and complex compounds even today has an advantage over other physical methods in some particular areas, e.g. metallisation of solids under high pressure, the nature of hydrogen bonds, or mutual influence of atoms in coordination compounds.

In the present work, we briefly summarise the physical foundations and structural applications of refractometry, the methods and results of measurements of RIs in elementary solids, binary and ternary inorganic compounds, complex (coordination) and organic crystalline substances. Extensive crystallo-optical studies, especially in the area of coordination compounds, were carried out by Soviet (Russian) scientists whose results are little known in the West, and this book has also the purpose of rectifying this deficiency. Unlike other available handbooks, this one pays attention to the effects of particle sizes, of pressure and temperature on the RIs of solids, including physical aftereffects in the structure and properties of shocked substances as well as anomalous dispersion of light and optical homogeneity in mixtures and solid solutions. Besides traditional techniques of RI measurements, we describe our development of the immersion method to enable studying highly refractive powder substances.

The earlier tables of RIs [1–5] listed also the densities and crystallographic parameters of the materials. We believe this is no longer necessary (except for substances previously unreported) because this information is readily available from structural databases and other online sources.

Regarding minerals, we give the data only for those of rational composition (daltonides), because optical properties of solid solutions usually can be calculated by additivity. For all RIs listed in this book, we provide references, except those taken from above-mentioned reference sources [1–5] or measured by ourselves and not yet published. The book consists of four chapters, dealing, respectively, with the physical theory, methods and results of RI measurements of various solids, and scientific and technological applications of these results.

Mendeleyevo, Russia   Stepan S. Batsanov

References

1. A.N. Winchell, H. Winchell, *Optical mineralogy* (New York, 1951)
2. E. Kordes, *Optische Data* (Verlag Chemie, Weinheim, 1960)
3. A.N. Winchell, H. Winchell, *Optical properties of artificial minerals* (Academic Press, New York and London, 1964)
4. M. Bass (ed.), *Handbook of optics*, 2nd edn., vol. 2 (McGraw-Hill, New York, 1995)
5. R.D. Shannon, R.C. Shannon, O. Medenbach, R.X. Fischer, J. Phys. Chem. Ref. Data, **31**, 931 (2002)

Contents

Part I Physical Definitions, Measurements and Applications of Refractive Indices

1	**Anisotropy, Dispersion, Theory and Structural Effects**	3
	References	7
2	**Methods of Measuring Refractive Indices**	9
	2.1 Method of the Prism	9
	2.2 Method of the Critical Angle	10
	2.3 Interferometric and Diffraction Methods	11
	2.4 Immersion Method	11
	2.5 Optical Homogeneity	13
	References	14
3	**Chemical Bonding and Refractive Indices**	17
	3.1 Density and the Refractive Index. Refraction	17
	3.2 Effects of Temperature and Pressure on the Refractive Index	24
	3.3 Effect of Grain Sizes in Solids on Their Refractive Indices	27
	References	28

Part II Anhydrous Substances

4	**Refractive Indices of Elements and Binary Compounds**	33
	References	40
5	**Refractive Indices of Ternary or Complex Halides and Oxides**	43
	References	48
6	**Refractive Indices of Silicates and Germanates**	51
	References	55
7	**Refractive Indices of Uranium Compounds**	57
	References	60

8   Refractive Indices of Oxygen-Containing Salts   61
      References.   68

**9   Refractive Indices in the Coordination Compounds
      of Group 11–14 Metals**   71
      References.   73

**10   Refractive Indices of Coordination Compounds
      of *d*- and *f*-Metals**   75
      References.   80

Part III   Crystallohydrates of Simple and Complex Compounds

11   Crystallohydrates of Simple and Complex Compounds   85
      References.   97

Part IV   Refractive Indices of Selected Organic Compounds

12   Refractive Indices of Selected Organic Compounds   103
      References.   104

Conclusion   107

Abstract

This book highlights the basics of crystal optics methods and refractive index (RI) measurement techniques in various solids, as well as their scientific and technological applications. Besides conventional methods of RI measurements, it describes special techniques where the former are impractical, e.g. for highly refracting powders, solids with anomalous dispersion of light and colloids. The tables compile all available RI measurements for elementary solids, binary, ternary and coordination compounds, as well as some small-molecule and polymeric organic substances.

Keywords Crystal optics · Anisotropy of solids · Optical/structural refractometry · Size effect · RIs of anhydrous solids · Ternary halides · Ternary oxides · Silicates · Uranium compounds · Organic substances

Part I
Physical Definitions, Measurements and Applications of Refractive Indices

Chapter 1
Anisotropy, Dispersion, Theory and Structural Effects

A ray of light is refracted when it crosses an interface between two media in which its *phase* velocities (v) are different. The refractive index (RI, n) which relates the angles (θ) of the incident (i) and refracted (r) rays is equal to the ratio of these velocities and hence is constant for any given pair of media.

$$n = \sin \theta_i / \sin \theta_r = v_i/v_r \qquad (1.1)$$

If one of the media is vacuum, where $v = c = 2.998 \times 10^8$ m/s, we measure *the absolute RI*:

$$n = c/v \qquad (1.2)$$

Usually, RI is measured in relation to atmospheric air, but since the latter has $n = 1.00027 \approx 1$, the measurements are close enough to the absolute RI for all chemical purposes. Gases, liquids, glasses and other non-crystalline (amorphous) solids,[1] as well as crystals of cubic symmetry are optically isotropic: v (and hence n) is equal in all directions. Light spreading from a point source in such a medium will have a spherical wave surface. In all other crystals, optical properties depend on the crystallographic direction. A light ray entering a crystal of hexagonal, trigonal or tetragonal symmetry splits into two, one of which ('ordinary' ray) has identical velocity, v_o (and accordingly identical RI, n_o), in all directions while the other ('extraordinary' ray) has direction-dependent velocity, v_e (and accordingly n_e). The former ray produces a spherical wave surface, and the latter that of a rotation ellipsoid whose axis, which coincides with the main crystallographic axis, is called the optic axis. Along this direction, the ordinary and extraordinary rays travel with

---

[1]Liquid crystals (liquids with partial ordering of molecules) and certain plastics under stress, are also optically anisotropic.

© The Author(s) 2016
S.S. Batsanov et al., *Refractive Indices of Solids*,
SpringerBriefs in Applied Sciences and Technology,
DOI 10.1007/978-981-10-0797-2_1

equal speed. Such crystals are called optically uniaxial. The difference $\Delta n = n_o - n_e$, often called birefringence, is the measure of optical anisotropy. Crystals with $n_o > n_e$ are known as optically negative, those with $n_o < n_e$ as positive. Finally, in crystals of orthorhombic, monoclinic and triclinic symmetry, a ray of light also splits into two components, but here *both* rays are 'extraordinary'; they are polarised in mutually perpendicular planes. The RI surface is an ellipsoid of a general type with three different principal half-axes, designated n_g, n_m and n_p (from the French *grand, moyen* and *petite*). In such ellipsoids, there are two directions (optical axes), the perpendicular cross-sections to which have the form of a circle; therefore, such crystals are named optically biaxial. Crystals with $n_g - n_m > n_m - n_p$ are known as optically positive, those with $n_g - n_m < n_m - n_p$ as negative. Note that optically uniaxial crystals can be described as biaxial with $n_o = n_m$ and $n_e = n_g/n_p$, depending on the optical sign of the crystal.

The average RI of a non-cubic crystal can be calculated by converting the RI ellipsoid into a sphere of equal volume,

$$\bar{n} = (n_e n_o^2)^{1/3} \quad \text{or} \quad \bar{n} = (n_g n_m n_p)^{1/3} \tag{1.3}$$

Refraction depends on the wave length; this property is known as (optical) dispersion. For uniformity, RI is commonly determined using the D line of sodium (589.3 nm or 2.10 eV) and denoted as n_D. The dispersion may be presented according to the Cauchy equation as

$$n = A + B/\lambda^2 + C/\lambda^4 \tag{1.4}$$

where A, B and C are empirical constants which can be determined by measuring RI at three different wave lengths. A equals RI when $\lambda = \infty$ and is known as n_∞. For approximate estimates of n_∞, only two constants are commonly used,

$$n = A + B/\lambda^2 \tag{1.5}$$

According to the electronic theory of Drude–Lorentz,

$$n^2 = 1 + \frac{N_1 e^2}{2\pi m} \Sigma \frac{C_i}{\omega_i^2 - \omega^2} \tag{1.6}$$

where N_1 is the particle density, e and m are the charge and the mass of the electron, C_i is the oscillator force, ω_i is the absorption frequency of the sample and ω is the frequency of the light used. It follows that the RI is the lowest at $\omega = 0$, i.e. at $\lambda = \infty$, and increases (together with the incident frequency) towards an absorption band, where $n \to \infty$ at $\omega \to \omega_i$. In the $\omega < \omega_i$ range, n increases as ω decreases (normal dispersion) but at $\omega > \omega_i$ the opposite is true (anomalous dispersion). For a single, main oscillator, Eq. 1.6 takes the form

$$n^2 = 1 + \frac{C_o}{E_o^2 - (\hbar\varpi)^2} \qquad (1.7)$$

where C_o is the oscillator force, E_o is the single oscillator energy, $\hbar\omega$ is the photon energy. At $\omega = 0$, Eq. 1.7 converts to a simpler form:

$$n^2 - 1 = C_o/E_o^2 \qquad (1.8)$$

Wemple and DiDomenico [1–3], using the well-known Kramers–Kronig relation, substituted C_o in Eq. 1.8 with the product E_oE_d where E_d is the dispersion energy and obtained

$$n^2 - 1 = \frac{E_d E_o}{E_o^2 - (\hbar\varpi)^2} \qquad (1.9)$$

Having analysed RI-dispersion data for over a hundred different solids and liquids, they established a simple empirical rule

$$E_d = \beta N_c N_e Z_a \qquad (1.10)$$

where N_c is the coordination number of the cation, Z_a is the formal charge of the anion, N_e is the effective number of valence electrons per anion (usually $N_e = 8$) and β is a constant equaling 0.26 ± 0.04 eV for ionic substances and 0.37 ± 0.05 eV for covalent ones. E_d is $ca.$ 1.5 times the width of the band gap (E_g), hence

$$(n^2 - 1)E_g = \frac{N_c}{a} N_e Z_a \qquad (1.11)$$

where $a = 4$ for covalent substances and $a = 6$ for ionic ones, i.e. the N_c/a ratio is the normalised (by the typical covalent or ionic values) coordination number. Substituting the n and E_g of vitreous As_2S_3, Se and Te into Eq. 1.11 gives for these amorphous materials the effective coordination numbers $N_c^* = 3.4$, 2.8 and 3.0, respectively, in good agreement with X-ray diffraction results. A sympathetic correlation of RIs with N_c in crystals has been established experimentally [4, 5] (see Table 1.1).

This dependence allows us to determine N_c in amorphous solids, glasses and films, where X-ray diffraction method is not very helpful. Thus, CaF_2 and Al_2O_3 epitaxy films have $n = 1.217$ and 1.632, respectively, which correspond to the structures with $N_c(Ca) = 6$ and $N_c(Al) = 4$ [4, 5]. Crystalline GeO_2 is converted to glass under shock compression; the heterogenous product was found to contain grains with $n = 1.608$–1.610 alongside those with $n = 1.8$–2.0, suggesting an increase of the N_c of Ge from 4 to 6 [6], which was later confirmed by static-compression experiments on vitreous GeO_2 [7, 8], similar to vitreous SiO_2 [9].

Table 1.1 Refractive indices in polymorphs with different coordination numbers

Crystal	N_c	n_D	Crystal	N_c	n_D	Crystal	N_c	n_D
HgS	2	2.29	GeO$_2$	4	1.708	CsBr	6	1.582
	4	3.37		6	2.016		8	1.698
C	3	2.03	Al$_2$O$_3$	4, 6	1.696	CsI	6	1.661
	4	2.42		6	1.766		8	1.788
BN	3	1.952	Eu$_2$O$_3$	6	1.983	SrF$_2$	8	1.435
	4	2.117		7	2.07		9	1.482
As$_2$O$_3$	3	1.755	Er$_2$O$_3$	6	1.953	BaF$_2$	8	1.475
	6	1.93		7	2.02		9	1.518
Sb$_2$O$_3$	3	2.087	Y$_2$O$_3$	6	1.915	PbF$_2$	8	1.766
	6	2.29		7	1.97		9	1.847
MnS	4	2.43	RbCl	6	1.51	ThF$_4$	8	1.530
	6	2.70		8	1.80		11	1.612
SiO$_2$	4	1.547	CsCl	6	1.534	UF$_4$	8	1.576
	6	1.812		8	1.642		11	1.685

Shannon et al. [10] analysed RIs of numerous oxides and fluorides using Eq. 1.7 and an alternative form of Sellmeier's equation,

$$1/(n^2 - 1) = -A/\lambda^2 + B \tag{1.12}$$

to calculate n_∞. They analysed the dispersion of RIs in terms of the Wemple–DiDomenico model (Eq. 1.9) and found several interesting effects. Thus, the effects of cation coordination were observed in Cu$_2$O, ZnO, arsenates, vanadates, iodates and molybdates, whereas all hydrates had a relatively low E_d. However, this model does not allow to calculate E_d for the majority of compounds because of the uncertainty in estimating N_e for s^2 and certain d^{10} compounds, or of the cation coordination where several cation types with different N_c are present simultaneously. These uncertainties, added to those concerning N_e and N_c, make it difficult to calculate E_d values in the majority of multi-ion compounds. At the same time, several empirical dependences of n on E_g are known [11–19]. All these formulae give qualitatively the same result: RI increases when E_g decreases. Because decreasing E_g amounts to increasing metallic character of bonding and in metals $n \to \infty$, this conclusion seems obvious. The RIs of MX compounds (M = Zn, Cd) decrease steadily in the succession MO < MS < MSe < MTe, while E_g increases, due to increasingly covalent character of Zn–X and Cd–X bonds in this succession [20]. However, it is not always so, viz. $n = 1.94$ for ZnO but 2.30 for ZnS, even though they have $E_g = 3.4$ and 3.9 eV, respectively.

It has been shown [21] that the observed RIs of TiOF$_2$, TiF$_4$ and of seven polymorphs of TiO$_2$ (columbite, rutile, brookite, anatase, ramsdellite, bronze and hollandite) can be explained not by their E_g but rather by the total absorption power per unit of volume

$$I(\varepsilon) = \frac{2}{\pi} \int\limits_{0}^{\infty} \varepsilon(\omega)\, \mathrm{d}\omega \qquad (1.13)$$

where ε is the dielectric permittivity. Thus for the same chemical composition, RI is inversely proportional to the cell volume per formula unit.

References

1. S.H. Wemple, M. Di Domenico, Phys. Rev. Lett. **23**, 1156 (1969)
2. S.H. Wemple, M. Di Domenico, Phys. Rev. B **3**, 1338 (1971)
3. S.H. Wemple, Phys. Rev. B **7**, 3767 (1973)
4. S.S. Batsanov, *Refractometry and chemical structure* (Van Nostrand, Princeton, 1966)
5. S.S. Batsanov, *Structural refractometry*, 2nd edn. (High School Press, Moscow, 1976). (in Russian)
6. S.S. Batsanov, E.V. Lazareva, L.I. Kopaneva, Russ. J. Inorg. Chem. **23**, 964 (1978)
7. J.P. Itie, A. Polian, G. Galas et al., Phys. Rev. Lett. **63**, 398 (1989)
8. Q. Mei, S. Sinogeikin, G. Shen et al., Phys. Rev. B **81**, 174113 (2010)
9. T. Sato, N. Funamori, Phys. Rev. B **82**, 184102 (2010)
10. R.D. Shannon, R.C. Shannon, O. Medenbach, R.X. Fischer, J. Phys. Chem. Ref. Data **31**, 931 (2002)
11. T.S. Moss, Proc. Phys. Soc. B **63**, 167 (1950)
12. T.S. Moss, Phys. Stat. Solidi B **131**, 415 (1985)
13. G. Dionne, J.C. Wooley, Phys. Rev. B **6**, 3898 (1972)
14. N. Ravindra, S. Auluck, V. Srivastava, Phys. Stat. Solidi B **93**, K155 (1979)
15. N.M. Ravindra, P. Ganapathy, J. Choi, Infrared Phys. Techn. **50**, 21 (2007)
16. A.L. Ruoff, Mater. Res. Soc. Symp. Proc. **22**, 279 (1984)
17. P. Herve, L.K.J. Vandamme, Infrared Phys. Technol. **4**, 609 (1994)
18. V. Dimitrov, S. Sakka, J. Appl. Phys. **79**, 1741 (1996)
19. S.K. Tripathy, Opt. Mater. **46**, 240 (2015)
20. X. Rocquefelte, M.-H. Whangbo, S. Jobic, Inorg. Chem. **44**, 3594 (2005)
21. X. Rocquefelte, F. Goubin, Y. Montardi et al., Inorg. Chem. **44**, 3589 (2005)

Chapter 2
Methods of Measuring Refractive Indices

Several methods for the RI measurement are available. RIs of bulk crystals and glasses can be measured very precisely using prisms cut-out of these materials [1–7] by the interference methods using a Michelson-type interferometer [8–10] or by the ellipsometer methods [11, 12]. RIs of solid *powders* are usually measured by the immersion method [13–15], whereas the optical diffraction method [16–18] is suitable for powders suspended in the liquid media. The authors' own works in this field are listed below.

2.1 Method of the Prism

A ray of light is directed on to a face of a prism at an angle of incidence θ_i, in such a way that the ray is parallel to the principal cross-section of the prism. The ray is refracted at the entrance face and then deviates from the original direction by an angle δ, forming an angle β with the normal to the exit face (exit angle) so that

$$\delta = \theta_i + \beta - \alpha \tag{2.1}$$

where α is the angle between the faces of the prism. Then RI can be calculated [4] as

$$n = \sqrt{\sin^2 \theta_i + (\sin \beta + \cos \alpha \cdot \sin \theta_i)^2 / \sin^2 \alpha} \tag{2.2}$$

In practice, one of the three angles (θ_i, β or δ) is usually fixed and only two have to be measured. The most common method is Fraunhofer's method of minimum deviation. Here, the prism is positioned so that the deviation angle δ reaches the minimum value (δ_o), the incident and exit angles are equal, and Eq. 2.2 is simplified to

© The Author(s) 2016
S.S. Batsanov et al., *Refractive Indices of Solids*,
SpringerBriefs in Applied Sciences and Technology,
DOI 10.1007/978-981-10-0797-2_2

$$n = \frac{\sin \frac{1}{2}(\delta_0 + \alpha)}{\sin \frac{\alpha}{2}} \tag{2.3}$$

Alternatively, the θ_i angle is fixed at $90°$, the light source being placed in the plane of the entrance face. With this geometry, one can observe the border between the illuminated and the dark fields, which corresponds to the complete internal reflection of the ray from the entrance face. Equation 2.2 then converts into

$$n = \sqrt{1 + \left(\frac{\cos \alpha + \sin \beta}{\sin \alpha}\right)^2} \tag{2.4}$$

The usual precision of RI measurements is $\pm 1.5 \times 10^{-5}$ using standard goniometers with 2–5″ angle readings. It can be increased to 10^{-6}, although this requires special goniometers with 0.1″ angle readings, careful thermostating and puts additional demands on the size and quality of the sample. The same method can be applied to measure RI in the UV and IR ranges adjacent to the visible range, using fluorescent eyepieces [8] or image-transforming tubes [9]. Measurements at non-ambient temperatures (−70 to +90 °C) are possible, albeit with reduced precision (several units of 10^{-4}). Various modifications of the prism method have been described [19]. A disadvantage of this method is the need to prepare prism-shaped samples, which is unsuitable for routine measurements.

2.2 Method of the Critical Angle

For the light to pass from a medium with higher RI to that with lower RI, the condition

$$\sin \theta_r = \frac{n_i}{n_r} \sin \theta_i < 1 \tag{2.5}$$

must be met. If θ_i exceeds the critical angle (corresponding to $\sin \theta_r = 1$) the light will experience total internal reflection (TIR) back into the high-RI medium. In this method, the sample is contacted to a reference prism with precisely known (high) RI and the critical angle, and hence, the sample RI is determined. It is impossible to measure RI higher than that of the measuring prisms, which for a long time were made of heavy flint glass ($n = 1.78$), but the introduction of thallium halide alloys or high-RI composite materials raised this limit to 2.00–2.20. Using photoelectric measuring device [19], RI can be measured with the precision of 10^{-5} in the visible and near-IR ranges.

Based on TIR of incoherent illumination, Abbé refractometers usually measure RIs of liquids but can also be used for solids with flat surfaces [20, 21]. Although TIR methods give good precision (10^{-4}–10^{-5}) for transparent liquids and solids, a small loss in the sample or incomplete contact between the sample and the prism severely increases the uncertainty in the critical angle and in the resulting RI value [21].

2.3 Interferometric and Diffraction Methods

On the contrary, for the interferometric methods, the accuracy in the RI measurement is not affected by the small loss in the sample, since the actual path length difference is measured. Although these methods are usually employed to detect very small relative phase differences, they were extended to measuring absolute RI values. A standard Michelson interferometer (MI) is used to measure such values in transparent solid plates, because of its simplicity [22, 23]. In the MI method, a transparent plate sample is rotated in one of the two arms of the interferometer continuously changing the optical path length difference, and hence producing a fringe pattern (with respect to the angle of incidence). From this pattern, and knowing the sample thickness, RI can be easily determined with the accuracy of $\sim 10^{-3}$, the error arising mainly from the sample thickness measurement, because the RI and the thickness cannot be independently determined from a single fringe pattern.

The Fabry–Perot (FP) method is another interferometric technique used for the same purpose [24]. In this case, the fringe pattern is determined solely by the phase difference between the directly transmitted light wave and the collinearly propagating waves, therefore the FP fringe pattern is more stable against environmental perturbations than the MI fringes. However, in both methods, the thickness measurement limits the accuracy of the RI determination. Gillen and Guha combined the MI and FP methods and successfully determined both the RI and the thickness values from the two correlated sets of fringes [25]. Coppola et al. [26] also applied FP method to obtain both RI and the thickness with the relative uncertainty of *ca.* 10^{-4}. A recent modification of the FP method [27] used two lasers with considerably different wavelengths, to measure the thickness and the RI with a relative uncertainty of 10^{-5}. We used the FP method to determine the RI of agglomerated nanodiamond particles in water colloidal solutions, having measured the particle sizes by dynamic laser scattering [28]. For different sizes of agglomerates, RI varied from 1.74 to 1.79, showing the agglomerates to consist of diamond ($n = 2.42$) and water ($n = 1.33$), presumably with water shells of 0.5 nm thickness surrounding 5 nm diamond particles [28].

2.4 Immersion Method

The immersion method, commonly used to determine RIs of polycrystalline materials, is based on the phenomenon of 'optical dissolution', whereby a crystalline grain becomes invisible in a liquid with the same RI. The right liquid can be chosen using the following effect. When a heterogeneous system, i.e. crystal grains in a liquid, is viewed under microscope in transmitted light, a bright rim (the Becke line) is visible around the phase boundaries. If the focal distance is increased, the Becke line always shifts towards the phase with higher RI, i.e. towards the centre of the particle if it has higher RI than the medium and vice versa.

Immersion liquids ('oils') with precisely known RI are commercially available. Standard sets have RIs ranging from 1.400 to 1.780, in 0.003–0.005 increments, and there are special compositions with higher RI, viz. S-As_2S_2-$AsBr_3$ (1:1:3) with $n = 2.00$, S-As_2S_3-$AsBr_3$ (1:7:12) with $n = 2.07$ and Se-As_2S_2-$AsBr_3$ (1:1:3) with $n = 2.11$. Interim values of RI can be obtained by adding methylene iodide (CH_2I_2) with $n = 1.74$ to standard oils. 'Fine-tuning' can be achieved by varying the temperature, because on warming RIs of liquids decrease much faster than those of crystals; so a crystal can be immersed into a liquid with higher RI and the whole warmed until the point of optical dissolution is reached.

Determination of higher RI presents considerable difficulties. Alloys of sulphur with selenium (n from 2.07 to 2.70) or thallium halides (TlCl $n = 2.25$, TlBr $n = 2.42$, TlI $n = 2.78$) can be heated to a plastic state and used as immersion media for measuring very high RIs. If heating affects the sample (through thermal decomposition or chemical interaction with the alloy), cold compression of the alloy powder and the sample into thin transparent disks can be used instead [29, 30]. For this purpose, we also prepared solid solutions TlCl/TlBr and TlBr/TlI with the content of each component varying from 0 to 100 % in 10 % steps. The RI of these materials was calibrated by using them as immersion media for measuring the *already known* RI of some crystals. This allowed us to measure high RIs with the accuracy of ± 0.01.

It is known that in IR spectra of powders, the intensity of absorption bands and of diffuse scattering depends on the difference between the RIs of the sample and the medium into which it is compressed [31]. To account for these effects, we studied [32] the IR spectra of several substances with known RIs compressed into transparent tablets with KBr, TlCl and TlBr and elucidated the dependence of the intensity of diffuse scattering on the difference of RIs in the samples and the immersion medium, Δn_{s-m}. Using these dependences, we determined for the first time the RIs of the powders of Mn_2O_3 (2.33), γ-MnS (2.45), α-MnS (2.67), MnSeS (2.79), MnSe (3.12), SnO (2.78), SnOS (2.67), $Sn_5O_6Se_4$ (2.75), $Sn_2O_3I_2$ (2.36) and $PtCl_2$ (2.24). These values are the exact RIs for optically isotropic and the average RIs for optically anisotropic materials.

The dependence of the intensity and shape of the IR valence vibration band v_3(N–O) at 1400 cm^{-1} on the medium has been studied on optically isotropic crystals of $Sr(NO_3)_2$ and $Pb(NO_3)_2$ [33] pressed into powders of KBr, KCl, CsCl, CsI, AgCl, CuCl, TlCl and TlBr as the immersion media with the RIs ranging from 1.490 to 2.302. It was found that when Δn_{s-m} decreases, the diffuse background in IR spectra decreases but the intensity of absorption bands and their fine structures improve. Because, in an absorption band, the anomalous dispersion occurs and the RI of a substance increases, it may become equal to that of the immersion medium. At this point, the mixture becomes transparent, i.e. the system acts like a Christiansen optical filter. Then, fixing the frequency at which the maximum transparency occurs and knowing the RI of the immersion medium at this frequency, the anomalous dispersion of the sample can be investigated, as has been done for $Sr(NO_3)_2$. In principle, using as the immersion medium a liquid with absorption bands in the visible range (where $n \rightarrow \infty$), one can determine the RI of any high-RI substance by varying the wavelength and measuring the light scattering in this liquid.

The dependence of the intensity of $v(C–O)$ absorption bands on Δn_{s-m} has been studied [34]. For optically uniaxial crystals of $MgCO_3$, $CaCO_3$ and $PbCO_3$, it revealed two maxima, corresponding to the two principal RIs, n_e and n_o; for optically biaxial $KHCO_3$, there were three maxima corresponding to three principal RIs (n_g, n_m, n_p). In the crystals of NH_4SCN and $Y(OH)_3$, in which quasi-isotropic ammonium cation and YO_6 polyhedron are combined with anisotropic SCN and OH ions, the intensities of the $v(N–H)$ and $v(Y–O)$ bond vibrations show 'isotropic' dependence on Δn_{s-m}, while the vibrations of SCN and OH follow the 'anisotropic' trend [35].

For an optically anisotropic crystal (see Chap. 1), RI depends on the direction, i.e. aspect of the crystal, and for a given aspect, on the direction in which the light is polarised. In optically uniaxial crystals, the ordinary ray (governed by n_o) is always polarised perpendicular to the optic axis, and the extraordinary ray (governed by n_e) in the direction parallel to the optic axis. Thus, n_o can be measured in any aspect, but n_e only in the plane parallel to the optic (=main crystallographic) axis, the chances of finding which incidentally are slim. In an arbitrary aspect, one would observe n'_e instead, which can vary from n_e to n_o. The standard routine is to inspect a sufficient number of variously oriented crystal grains, find the limit of n'_e and assume it to be n_e. Similarly, for a random aspect of a biaxial crystal, one can measure two RIs, n'_g and n'_p, related to the principal RIs of the crystal as $n_g > n'_g > n_m > n'_p > n_p$. Note that the sufficient number of observations [36] should be at least equal to the birefringence Δn divided by the precision of the immersion method, i.e. usually $\Delta n/0.002$ for uniaxial and $\Delta n/0.003$ for biaxial crystals [36]. Hence, a reliable characterisation of a material may require tens or even hundreds of observations, even assuming fully random grain orientations. The latter is unlikely, as optically anisotropic grains tend to have also anisotropic shapes and mechanical properties, thus making the task even more difficult.

2.5 Optical Homogeneity

If a solid is heterogeneous but the grain sizes of the components are smaller than the wavelength of light, for crystal-optical investigation, it will appear as homogeneous with a uniform RI – although other methods, such as X-ray diffraction, may recognise the presence of different phases. The phenomenon was first described by Belyankin [37] with respect to nuclei of the mineral mullite formed within a glass and was extensively explored since then [38–50]. Over 400 cases have been described so far of this effect, known as 'optical homogeneity' in Western literature, while the term 'optical mixing' [38] is used in Russian. Obviously, it becomes increasingly relevant with the current intense interest in sub-micron and nano-size powders. A similar effect was observed in solid products of shock compression, when partial phase transformations took place under certain thermodynamic conditions [50].

The usual sources of optically homogeneous systems are as follows: (i) high purity solid-phase processes which do not involve melting, (ii) chemical transformations with melting of initial components and the reaction products and (iii) physical transformations at temperatures above the melting point. It can occur both when a crystal undergoes a partial amorphisation (on heating) and when an amorphous solid or gel undergoes partial crystallisation, as well as during a concomitant or consequent crystallisation of different phases, one of them crystallising on the surface of another. In fact, the effect can be modelled by deliberately mixing fine-grain components. This has important implications for the immersion method, where a sample is always ground up prior to investigation, and optical homogenisation may occur at this stage.

It is also important that crystallisation of a new phase, microscopically intergrown with the starting material, is always oriented in certain directions with respect to the latter, in accordance with the elements of structural similarity between the two. Such intergrowth typically results in optical homogeneity. Thus, we observed a pseudomorphic optically homogeneous phase during crystallisation of alkali aluminates of the β-alumina type, when the structures of both the initial Θ-Al_2O_3 and the product have common layered blocks of a layered spinel type. An optically homogeneous phase was observed in the products of the interaction between β-eucryptite and dolomite, where prismatic crystals with an RI intermediate between those of γ-$LiAlO_2$ and β-Ca_2SiO_4, were observed.

In hydrothermal systems, we have observed generation of optically homogeneous phases in various forms, e.g. needle-shaped. The most probable reason for the formation of metastable combined phases seems to be the existence of similar structural elements in the two phases. The duration of optically homogeneous phase existence depends on temperature, the degree of structural likeness and the ability of the more disperse phase to crystallise. The available data suggest that optically homogeneous phases can be much more widespread than commonly recognised, and may emerge in any method of synthesis, even simple mechanical grinding. On the other hand, the existence of such phases can provide insight into the structural elements of the intergrown components.

References

1. L.W. Tilton, J. Res. NBS **14**, 393 (1935)
2. L.W. Tilton, E.K. Plyler, R.E. Stephens, J. Res. NBS **43**, 81 (1949)
3. A.J. Werner, Appl. Opt. **7**, 837 (1968)

4. B.V. Ioffe, *Refractometrical methods in chemistry*, 3rd edn. (Khimia, Leningrad, 1983). (in Russian)
5. D. Tentoriand, J.R. Lerma, Opt. Eng. **29**, 160 (1990)
6. O. Medenbach, R.D. Shannon, J. Opt. Soc. Am. B **14**, 3299 (1997)
7. O. Medenbach, D. Dettmar, R.D. Shannon et al., J. Opt. A: Pure Appl. Opt. **3**, 174 (2001)
8. J. Grehn, Leitz-Mitt. Wiss. Technol. **1**, 35 (1959)
9. M.S. Shumate, Appl. Opt. **5**, 327 (1966)
10. V.A. Moskalev, L.A. Smirnova, Sov. J. Opt. Technol. **54**, 461 (1987)
11. G.E. Jellison, F.A. Modine, Appl. Opt. **36**, 8184 (1997)
12. G.E. Jellison, ibid, 8190
13. C.P. Saylor, J. Res. NBS **14**, 277 (1935)
14. A.M. Kauffman, Thin Solid Films **1**, 131 (1967)
15. D.J. Little, D.M. Kane, Opt. Express **19**, 19182 (2011)
16. T. Kinoshita, Adv. Powder Technol. **12**, 589 (2001)
17. E. Pol, F.A.W. Coumans, A. Stutk et al., Nano Lett. **14**, 6195 (2014)
18. C. Meichner, A.E. Schedl, C. Neuber et al., AIP Adv. **5**, 087135 (2015)
19. B.I. Molochnikov, B.Ya. Karasik, M.V. Laikin, Optico-mechanic industry, **7**, 36 (1974). (in Russian)
20. H. Onodera, I. Awai, J. Ikenoue, Appl. Opt. **22**, 1194 (1983)
21. G.H. Meeten, Measur. Sci. Technol. **8**, 728 (1997)
22. M.S. Shumate, Appl. Opt. **5**, 327 (1966)
23. G.D. Gillen, S. Guha, Appl. Opt. **43**, 2054 (2004)
24. J.C. Brasunas, G.M. Curshman, Opt. Eng. **34**, 2126 (1995)
25. G.D. Gillen, S. Guha, Appl. Opt. **44**, 344 (2005)
26. G. Coppola, P. Ferraro, M. Iodice, S. De Nicola, Appl. Opt. **42**, 3882 (2003)
27. H.J. Choi, H.H. Lim, H.S. Moon et al., Opt. Express **18**, 9429 (2010)
28. S.S. Batsanov, E.V. Lesnikov, D.A. Dan'kin, D.M. Balakhanov, Appl. Phys. Lett. **104**, 133105 (2014)
29. E.D. Ruchkin, S.S. Batsanov, Proc. Sibir. Div. Acad. Sci. USSR, No 11, 122 (1963) (in Russian)
30. E.D. Ruchkin, Yu.I. Vesnin, S.S. Batsanov, Crystallography **9**, 749 (1964). (in Russian)
31. S.E.F. Smallwood, P.B. Hart, Spectrochim. Acta **19**, 285 (1963)
32. S.S. Batsanov, Z.A. Grankina, Opt. Spectrosc. **19**, 814 (1965). (in Russian)
33. S.S. Batsanov, S.S. Derbeneva, Opt. Spectrosc. **17**, 149 (1964). (in Russian)
34. S.S. Batsanov, S.S. Derbeneva, Opt. Spectrosc. **18**, 599 (1965). (in Russian)
35. S.S. Batsanov, S.S. Derbeneva, Opt. Spectrosc. **22**, 157 (1967). (in Russian)
36. S.S. Batsanov, Bull. Moscow Univ. **4**, 127 (1958). (in Russian)
37. D.S. Belyankin, B.V. Ivanov, B.V. Lapin, *Petrography of technical rocks* (Acad. Sci. USSR Press, Moscow, 1952) (in Russian)
38. V.I. Muravjev, V.A. Drits, *Clays, their mineralogy and applications* (Nauka, Moscow, 1970), p. 53 (in Russian)
39. A.S. Marfunin, Doklady Acad. Sci. USSR **127**, 869 (1959) (in Russian)
40. V.A. Frank-Kamenetsky, *Nature of structural admixture in minerals* (Leningrad University Press, Leningrad, 1964)
41. J. Hauser, H. Wenk, Z. Krist. **143**, 188 (1976)
42. I.A. Poroshina, A.S. Berger, Proc. Miner. Soc. USSR **105**, 369 (1976). (in Russian)
43. S.S. Batsanov, I.A. Poroshina, Proc. Miner. Soc. USSR **108**, 74 (1979). (in Russian)
44. I.A. Poroshina, M.I. Tatarintseva, Proc. Miner. Soc. USSR **109**, 728 (1980). (in Russian)
45. K.O. Dornberger-Shiff, G. Grell, Crystallography **27**, 126 (1982). (in Russian)
46. I.A. Poroshina, S.S. Batsanov, Proc. Miner. Soc. USSR **117**, 212 (1988). (in Russian)
47. G.A. Lager, Th. Armbruster, D. Pohl, Phys. Chem. Miner. **9**, 177 (1997)
48. B.B. Shkursky, Proc. Educat. Instit. Geol. **4**, 37 (2005)
49. B.B. Shkursky, Proc. Educat. Instit. **3**, 22 (2008)
50. E.D. Ruchkin, M.N. Sokolova, S.S. Batsanov, J. Struct. Chem. **8**, 410 (1967)

Chapter 3
Chemical Bonding and Refractive Indices

Any structural change affecting the density and chemical bonding in a substance will also affect its RI. Therefore, knowledge of RIs is very useful for understanding electronic structure of materials, especially in the case of fine powders, amorphous solids and glasses, which are difficult for X-ray diffraction methods.

3.1 Density and the Refractive Index. Refraction

Newton was the first to notice, in his famous *Opticks* (1704) [1], that RI is correlated with density. He observed that for many substances, ranging from atmospheric air to diamond, the parameter (later called 'specific refraction')

$$r_1 = (n^2 - 1)\rho \qquad (3.1)$$

varies only by a factor of 3, while the density varies by three orders of magnitude. In 1853, Beer [2] suggested that for gases the expression

$$r_2 = (n - 1)/\rho \qquad (3.2)$$

is more invariant with respect to thermodynamic conditions. Later, Gladstone and Dale [3] concluded that the same is true for liquids and showed that r_2 is only slightly affected by changes of temperature and the aggregate state of a substance, by mixing with other liquids, or even, within certain limits, by a chemical interaction. In 1875–1880, the formula

$$\frac{n^2 - 1}{n^2 + 2} = r_3\rho \qquad (3.3)$$

© The Author(s) 2016
S.S. Batsanov et al., *Refractive Indices of Solids*,
SpringerBriefs in Applied Sciences and Technology,
DOI 10.1007/978-981-10-0797-2_3

was derived by Lorenz [4, 5] from his own theory of the propagation of light (a precursor of Maxwell's theory) and from the classical electromagnetic theory of Lorentz [6, 7]. Other formulae have been also suggested [8, 9]. For non-critical gases, where $n \approx 1$ and intermolecular interactions are slight, Eqs. 3.1–3.3 give similar results (with $r_1 \approx 2r_2 \approx 3r_3$). Essential differences appear for condensed phases, where each molecule is substantially polarised by its neighbours. For these, Eq. 3.3 is the most rigorous, related as it is to the simple but efficient model of such polarisation by Clausius and Mossotti (see below), especially if instead of density (specific gravity) ρ we use *particle* density, i.e. the inverse of molar volume.

A product of the molecular mass (M) and r is called molar refraction (R) and is most commonly used in structural chemistry, in the form defined by the Lorenz–Lorentz equation (see details in [8, 9])

$$R = V \frac{n^2 - 1}{n^2 + 2} \tag{3.4}$$

Molar refraction can be approximated by a sum of constant increments attributed to specific atoms, ions and/or chemical bonds present in the molecule. Comparison of this additive R with the observed value can yield useful information about the structure and bonding.

The covalent refractions of atoms in A_2 molecules are close to the atomic increments derived by Vogel and Miller from experimental molecular refractions of organic compounds. These results are presented in Table 3.1. The increments per double and triple bonds involving carbon reflect an increase of $R(C)$ upon a decrease of N_c (cf. Table 3.2), while different refractions of the same atom in

Table 3.1 Atomic refractions (cm³/mol) of Vogel[a] and Miller[b]

Atoms, groups	R_D^a	R_D^b	R_∞	Atoms, groups	R_D^a	R_∞
H	1.03	0.98	1.01	CN	5.46[c]	5.33
C	2.59	2.68	2.54	NO₃	9.03	8.73
O	1.76	1.61	1.72	CO₃	7.70	7.51
OH	2.55	2.58	2.49	SO₃	11.34	11.04
F	0.81	0.75	0.76	SO₄	11.09	10.92
Cl	5.84	5.84	5.70	PO₄	10.77	10.63
Br	8.74	7.60	8.44	CH₂	4.65	4.54
I	13.95	13.66	13.27	CH₃	5.65	5.54
N (aliphatic)	2.74	2.43	2.57	Formation of:		
N(aromatic)	4.24	2.75	3.55	3-member cycle	0.60	0.53
—ONO (nitrito)	7.24		6.95	4 member cycle	0.32	0.28
—NO₂ (nitro)	6.71		6.47	5-member cycle	−0.19	−0.19
S	7.92	7.57	7.60	6-member cycle	−0.15	−0.15
SCN (thiocyano)	13.40		12.98	Double bond	1.58[d]	1.42
NCS (isothiocyano)	15.62		14.85	Triple bond	1.98[e]	1.85

[a][61, 62]; [b][63]; [c]5.65 [63]; [d]1.47 [63]; [e]1.12 cm³/mol. [63]

Table 3.2 Molar refractions R_D (cm^3/mol) of polymorphs with different coordination numbers [10, 64]

Substance	Polymorph	N_c	R_D	Substance	Polymorph	N_c	R_D
C	Diamond	4	2.11	Y_2O_3	Hexagonal	6	21.2
	Graphite	3	2.70		Cubic	7	20.5
BN	Diamond	4	3.83	Nd_2O_3	Hexagonal	6	25.5
	Graphite	3	5.27		Cubic	7	24.2
MnS	B1	6	14.5	Eu_2O_3	Hexagonal	6	23.9
	B3	4	16.3		Cubic	7	23.1
CsCl	B2	8	15.20	Gd_2O_3	Hexagonal	6	23.5
	B1	6	16.15		Cubic	7	22.7
CsBr	B2	8	18.44	Dy_2O_3	Hexagonal	6	22.3
	B1	6	18.84		Cubic	7	21.5
CsI	B2	8	24.19	Ho_2O_3	Hexagonal	6	21.7
	B1	6	25.01		Cubic	7	21.1
SrF$_2$	Cottunite	9	7.63	Er_2O_3	Hexagonal	6	21.3
	Fluorite	8	7.78		Cubic	7	20.7
BaF$_2$	Cottunite	9	9.78	Tu_2O_3	Hexagonal	6	21.0
	Fluorite	8	10.09		Cubic	7	20.2
PbF$_2$	Cottunite	9	12.94	Yb_2O_3	Hexagonal	6	20.6
	Fluorite	8	13.08		Cubic	7	19.7
SiO$_2$	Rutile	6	6.01	Lu_2O_3	Hexagonal	6	20.0
	Quartz	4	7.19		Cubic	7	19.3
GeO$_2$	Tetragonal	6	8.47	Al_2O_3	α-phase	6	10.6
	Hexagonal	4	9.53		γ-phase	4, 6	11.3

different functional groups reflect the effects of chemical bonding. Tables 3.3 and 3.4 list the bond refractions for organic, organometallic and coordination compounds, Table 3.5 gives a complete system of refractions of atoms in an isolated (free) state, in X_2 molecules and in elementary solids [10].

The best additive calculations of molecular refractions of organic compounds are obtained using only bond increments (no atomic ones!), see Table 3.3.

In coordination compounds, bond refractions are not additive, because each bond is strongly affected by its *trans* counterpart. Therefore, instead of bond increments, Yakshin introduced 'coordinate refractions' [11], i.e. combinations of bonds lying on the same Werner coordinate in a coordination polyhedron of the square or octahedral type. Thus, R(X–M–X) can be calculated as R(MX$_4$)/2 or R(MX$_6$)/3. Yakshin found also that the influence of ligands in a *cis*-position to a given atom on its coordinate refraction is negligible. This approach allowed us later to give a quantitative characteristic to the mutual influence of atoms in coordination compounds. The effect of transinfluence was discovered by Chernyaev in 1926 (for the history see [12–15]), and the coordinate refractions were the first physical characteristics of this effect [16–19]. The early history of the theory and physical

Table 3.3 Bond refractions (cm^3/mol) in the systems of Vogel[a] and Miller[b] [10]

Bond	R_D^a	R_∞^a	R_D^b	Bond	R_D^a	R_∞
C–H	1.676	1.644	1.645	C=O in MeC(O)R	3.49	3.38
C–C	1.296	1.254	1.339	C–S	4.61	4.42
C–C (cyclopropane)	1.50	1.44		C=S	11.91	10.79
C–C (cyclobutane)	1.38	1.32		C–N	1.57[c]	1.49
C–C (cyclopentane)	1.28	1.24		C=N	3.75	3.51
C–C (cyclohexane)	1.27	1.23		C≡N	4.82	4.70
C$_{ar}$–C$_{ar}$	2.69	2.55	2.74	N–N	1.99	1.80
C=C	4.17	3.94	4.14	N=N	4.12	3.97
C≡C	5.87	5.67	5.13	N–H	1.76[d]	1.74
C–F	1.55	1.53	1.40	N–O	2.43	2.35
C–Cl	6.51	6.36	6.51	N=O	4.00	3.80
C–Br	9.39	9.06	8.27	O–H (alcohols)	1.66[e]	1.63
C–I	14.61	13.92	14.33	O–H (acids)	1.80	1.78
C–O (ethers)	1.54	1.49	1.47	S–H	4.80	4.65
C–O (acetals)	1.46	1.43		S–O	4.94	4.75
C=O	3.32	3.24	2.57	S–S	8.11	7.72

Additions to Vogel's system

Bond	R_D	Bond	R_D	Bond	R_D
O–O	2.27	Si–Br	10.24	Sn–Sn	10.7
Se–Se	11.6	Si–O	1.80	B–H	2.15[f]
P–H	4.24	Si–S	6.14	B–F	1.68
P–F	3.56	Si–N	2.16	B–Cl	6.95
P–Cl	8.80	Si–C$_{alkyl}$	2.47	B–Br	9.6[f]
P–Br	11.64	Si–C$_{aryl}$	2.93	B–O	1.61
P–O	3.08	Si–Si	5.87	B–S	5.38
P–S	7.56	Ge–H	3.64	B–N	1.96
P = S	6.87	Ge–F	2.3	B–C$_{alkyl}$	2.03
P–N	2.82	Ge–Cl	7.65	B–C$_{aryl}$	3.07
P–C	3.68	Ge–Br	11.1	Al–O	2.15
As–O	4.02	Ge–I	16.7	Al–N	2.90
As–C	4.52	Ge–O	2.50	Al–C	3.94
As–Cl	9.23	Ge–S	7.02	Hg–Cl	7.63[f]
As–Br	13.3	Ge–N	2.33	Hg–Br	9.77[f]
As–I	20.4	Ge–C	3.05	Hg–C	7.21
Sb–H	3.2	Ge–Ge	6.85	Zn–C	5.4
Sb–Cl	10.6	Sn–H	4.83	Cd–C	7.2
Sb–Br	13.6	Sn–Cl	8.66	In–C	5.9
Sb–I	20.8	Sn–Br	11.97	Pb–C	5.25
Sb–O	5.0	Sn–I	17.41	Sb–C	5.4
Sb–C	5.4	Sn–O	3.84	Bi–C	6.9

(continued)

Table 3.3 (continued)

Si–H	3.0	Sn–S	7.63	Se–C	6.0	
Si–F	2.1	Sn–C_{alkyl}	4.17	Te–C	7.9	
Si–Cl	7.92	Sn–C_{aryl}	4.55			

[a][61, 62], [b][63]; [c]1.48 [63]; [d]1.79 [63]; [e]1.78 cm^3/mol [63]; [f]R_∞

Table 3.4 Coordination refractions (cm^3/mol) in coordination compounds of platinum, palladium and cobalt

Coordinate	R_∞	Coordinate	R_∞	Coordinate	R_∞	Coordinate	R_∞
Cl–Pt^{II}–Cl	18.93	Cl–Pt^{IV}–Cl	17.84	Cl–Pd^{II}–Cl	18.94	Cl–Co^{III}–Cl	17.82
Cl–Pt^{II}–NH$_3$	16.00	Cl–Pt^{IV}–NH$_3$	14.86	Cl–Pd^{II}–NH$_3$	15.72	Cl–Co^{III}–NH$_3$	13.01
NH$_3$–Pt^{II}–NH$_3$	12.65	NH$_3$–Pt^{IV}–NH$_3$	11.43	NH$_3$–Pd^{II}–NH$_3$	12.12	NH$_3$–Co^{III}–NH$_3$	10.61
NH$_3$–Pt^{II}–NO$_2$	17.33	NH$_3$–Pt^{IV}–NO$_2$	16.40	NH$_3$–Pd^{II}–NO$_2$	17.98	NH$_3$–Co^{III}–NO$_2$	15.31
NO$_2$–Pt^{II}–NO$_2$	21.16	NO$_2$–Pt^{IV}–NO$_2$	21.00	NO$_2$–Pd^{II}–NO$_2$	21.58	NO$_2$–Co^{III}–NO$_2$	20.58
CNS–Pt^{II}–SCN	34.88	NO$_2$–Pt^{IV}–Cl	19.94	NO$_2$–Pd^{II}–Cl	20.56		
CNS–Pt^{II}–NH$_3$	23.25	Br–Pt^{IV}–Br	24.12	CNS–Pd^{II}–SCN	34.34		
CN–Pt^{II}–NC	20.92			CN–Pt^{II}–NC	18.06		
Br–Pt^{II}–Br	23.66			NC–Pt^{II}–CN	17.10		

corollaries of transinfluence in coordination compounds of Group 8–10 metals are given in [8, 9]; for later studies, see [20, 21].

Table 3.5 shows that refractions of free metallic atoms are, as a rule, significantly higher than those of covalently bonded atoms. However, solid metals have very high RIs (see below, Table 4.1) and, hence, according to Eq. 3.4, $R \approx V$. According to the Goldhammer–Herzfeld criterion, $V \rightarrow R$ when a dielectric converts into a metal [22, 23]. Hence, the ratio R/V can be considered as a measure of bond metallicity, and the pressure at which $V = R$ is often regarded as the pressure of metallisation. The Goldhammer–Herzfeld criterion is not precisely correct, however, since the RIs of solids increase under any compression; although it is valid for rough estimates of the pressures of metallisation. A more rigorous approach was realised in high-pressure studies of CH$_4$ and SiH$_4$ [24], which showed sharp increases of the R/V ratios at 288 and 109 GPa, respectively, indicating phase transformations of the insulator–semiconductor type.

Molar refractions of polar inorganic compounds are usually calculated using ionic refractions; their recommended values are given in Table 3.6. Calculations of molar refractions in inorganic compounds with intermediate bonding character are discussed in Refs. [8–10, 25].

An important part of structural refractometry is the determination of atomic sizes. Because refractions of atoms are proportional to their electronic polarisability (α)

$$R = V \frac{n^2 - 1}{n^2 + 2} = \frac{4}{3} \pi N \alpha \tag{3.5}$$

Table 3.5 Refractions (cm^3) of atoms in free state (upper lines), in diatomic molecules (middle lines), in elementary solids (bottom lines); $R(H) = 1.68$ in the free state and 1.02 in H_2, R (He) = 0.52

Li	Be	B	C	N	O	F	Ne		
59.1	14.0	7.6	4.22	3.05	1.97	1.40	1.00		
41.4	10.8	4.3	2.07	2.20	1.99	1.45			
13.0	4.9	3.5	2.07						
Na	Mg	Al	Si	P	S	Cl	Ar		
60.2	27.5	21.9	13.9	9.2	7.3	5.5	4.14		
49.9	19.3	11.5	9.05	8.57	7.7	5.69			
23.6	14.0	10.0	9.05	8.75	7.7				
K	Ca	Sc	Ti	V	Cr	Mn	Fe	Co	Ni
110	61.6	44.9	36.8	31.3	29.3	23.7	21.2	18.9	17.1
93.1	46.2	32.5	27.3	21.1	20.9	19.5	14.0	13.3	12.5
45.6	26.3	15.0	10.6	8.4	7.2	7.3	7.1	6.7	6.6
Cu	Zn	Ga	Ge	As	Se	Br	Kr		
28.5	14.5	20.5	14.7	10.9	9.5	7.7	6.27		
8.1	12.9	17.2	11.3	10.9	10.8	8.17			
7.1	9.2	11.7	11.3	10.3	10.6	8.75			
Rb	Sr	Y	Zr	Nb	Mo	Tc	Ru	Rh	Pd
121	72.2	57.2	45.1	39.6	32.3	28.8	24.2	21.7	20.1
99.6	55.0	38.0	27.6	23.8	21.4	19.0	13.3	14.0	14.7
55.9	33.9	20.0	14.0	10.9	9.4	8.6	8.2	8.2	8.8
Ag	Cd	In	Sn	Sb	Te	I	Xe		
28.7	18.6	25.7	15.8	16.6	13.9	12.5	10.20		
11.9	14.5	21.6	16.3	17.7	14.4	13.0			
10.3	12.9	15.8	16.3	17.7	15.4				
Cs	Ba	La	Hf	Ta	W	Re	Os	Ir	Pt
150	95.6	78.4	40.9	33.0	28.0	24.5	21.4	19.2	16.4
131	71.8	55.3	26.7	16.7	15.6	15.5	11.2	11.0	10.0
69.7	37.9	22.5	13.5	10.8	9.6	8.8	8.4	8.5	9.1
Au	Hg	Tl	Pb	Bi	Th	U	Rn		
20.9	12.7	19.2	17.6	18.7	81.0	51.2	13.4		
11.0	12.8	16.7	18.4	21.3	57.3	38.2			
10.2	13.9	17.1	18.3	21.3	19.8	12.5			

and $\alpha = r^3$ (as required by the Clausius–Mossotti theory), refractometric data allow to calculate geometrical characteristics of ions and bonds in good agreement with crystallographic results [8, 9].

The applications of refractometry to organic molecules, silicates, geometrical isomers of coordination compounds, etc., are now only of historical interest, but for studies of hydrogen bonding and mutual influence of atoms in a molecule, this method is still relevant. Table 3.7 lists the major results of refractometric studies of

Table 3.6 Ionic refractions (cm^3/mol)

+1		+2				+3		+4	+5	-1^a	
Li	0.07	Be	0.02	Cr	1.8	B	0.01	Si 0.10	P	0.06	F	2.5
Na	0.45	Mg	0.25	Mn	1.6	Al	0.15	Ge 0.5	As	0.3		2.5
K	2.2	Ca	1.5	Fe	1.5	Ga	0.6	Sn 1.4	Sb	1.2	Cl	8.0
Rb	3.5	Sr	2.5	Co	1.4	In	1.8	Pb 1.8	Bi	1.5		8.5
Cs	6.2	Ba	4.6	Ni	1.3	Tl	2.0	Ti 1.0	V	0.8	Br	11
Cu	1.5	Zn	0.9	Ru	2.6	Sc	1.2	Zr 1.7	Nb	1.2		11.8
Ag	3.5	Cd	2.5	Rh	2.4	Y	2.0	Hf 1.6	Ta	1.1	I	17
Au	4.5	Hg	3.0	Pd	2.9	La	3.5	Cr 0.9	+6			18
Tl	10.0	Cu	1.1	Os	2.6	V	1.7	Mo 1.4	S	0.04	-2	
		Sn	8.0	Ir	2.5	As	4.5	W 1.4	Se	0.25	O	7.5
		Pb	9.0	Pt	2.4	Sb	6.5	Te 5.0	Te	0.8		8.0
						Bi	7.5	Mn 0.9	Cr	0.6	S	17
						Cr	1.6	Pt 1.4	Mo	1.2		18
						Mn	1.5	Th 4.5	W	1.2	Se	21
						Fe	1.4	U 4.0				22.5
						Co	1.3				Te	29
						Ni	1.2					31
						U	4.5					

^{a}upper lines: R_∞, lower lines: R_D

Table 3.7 Refractions (cm^3/mol) of hydrogen bonds [8, 9]

Acids	$R_{XH \cdots X}$	Acid salts	$R_{H \cdots X}$
HF	0.38	KHF_2	0.43
HNO_3	0.56	$KHCO_3$	0.35
H_2SO_4	0.70	$KHSO_4$	0.41
H_3PO_4	0.75	K_2HPO_4	0.53
$H_3Fe(CN)_6$	0.64	KH_2PO_4	0.66
$H_4Fe(CN)_6$	1.08		
Ammonium salts	$R_{NH \cdots X}$	Crystallohydrates	$R_{OH \cdots X}$
NH_4HF_2	0.09	$KF \cdot 2H_2O$	0.09
NH_4NO_3	0.07		
NH_4HCO_3	0.09	$Na_2CO_3 \cdot 10H_2O$	0.18
$(NH_4)_2SO_4$	0.09	$Na_2SO_4 \cdot 10H_2O$	0.18
$(NH_4)_2HPO_4$	0.13	$Na_2HPO_4 \cdot 12H_2O$	0.19
$NH_4H_2PO_4$	0.13	H_2O liquid	0.15
$(NH_4)_3Fe(CN)_6$	0.12	H_2O solid	0.16
$(NH_4)_4Fe(CN)_6$	0.29	$K_4Fe(CN)_6 \cdot 3H_2O$	0.27

hydrogen bonds in inorganic compounds, showing a change of the bond strengths in the succession acid > acid salts > ammonium salts > crystallohydrates. This sequence is caused by the variation in the effective atomic charges: an accumulation of hydrogen atoms in the outer sphere of a complex ion in acid salts or acids enhances the polarity of the oxygen or nitrogen atoms and hence the strength of H-bonds with the latter. The refraction of liquid water changing from 3.66 cm^3/mol at ambient conditions to 3.20 cm^3/mol under shock compression at $P = 22$ GPa [26] indicates a disruption of H-bonds in the latter [27].

3.2   Effects of Temperature and Pressure on the Refractive Index

Measurements of RIs of crystals on heating allow to study the nature of chemical bonds in real time. In solid dielectrics, including glasses, dn/dT depends only upon the temperature-induced change in density, ρ [28]:

$$\frac{\mathrm{d}n}{\mathrm{d}T} = \left(\frac{\partial n}{\partial T}\right)_\rho + \left(\frac{\partial n}{\partial \rho}\right)_T \left(\frac{\partial \rho}{\partial T}\right) \tag{3.6}$$

The factor $\delta = \frac{1}{n}\frac{\partial n}{\partial T}$ is different for ionic and semiconductor crystals (Table 3.8). In the former, RI decreases only due to decreasing density (thermal expansion), whereas in semiconductors there is a simultaneous increase of the polarisability due to an emergence of free electrons. Considering the expansion and electronic factors together, it was possible to classify materials on the basis of their δ; the averaged values of δ for elements and binary compounds [8, 29] are listed in Table 3.8. For anisotropic crystals, the values averaged δ according to Eq. 3.7 are presented, except for TiO$_2$, KNbO$_3$ and Tl$_3$AsSe$_3$ where δ has different signs in different crystallographic directions.

If the composition of a substance changes under heating, RI measurements help us to understand the features of their structures. Thus, elimination of water from crystalline hydrates alters their structures with an increasing of density and RI, but in zeolites, a dehydration without structural change reduces both ρ and RI [8, 9].

An increase of density in substances at high pressures, proportional to their compressibility, increases the number of electronic oscillators per unit volume and, hence, the RI,

$$\rho\left(\frac{\partial n}{\partial \rho}\right) = \frac{1}{B}\left(\frac{\mathrm{d}n}{\mathrm{d}P}\right) \tag{3.7}$$

where B is the bulk modulus. But at the same time, compression of the electron clouds of atoms reduces their polarisability and hence, RI. For the latter case Mueller [30] introduced a special factor

Table 3.8 Temperature coefficients (δ) of refractive indices in crystals

Substance	$-\delta \times 10^5$	Substance	$-\delta \times 10^5$	Substance	$+\delta \times 10^5$
LiYF$_4$	0.13	AgCl	6.1	LiNbO$_3$	0.9
LiF	1.2	CaF$_2$	1.0	BeO	1.0
NaF	1.1	SrF$_2$	1.6	MgO	1.65
NaCl	2.7	BaF$_2$	1.6	β-Ga$_2$O$_3^a$	7.32–7.61
NaBr	3.5	PbI$_2$	8.0	Diamond	1.0
NaI	4.4	Ar	27	Si	16.6
KF	2.2	Kr	32	Ge	46.2
KCl	2.8	Xe	29	ZnS	6.3
KBr	3.3	PbS	210	ZnSe	9.1
KI	3.0	PbSe	230	CdS	6.0
RbCl[b]	4.1	PbTe	210	CdTe	14.7
RbBr[b]	6.7	SiO$_2$	0.4	GaN	6.1
CsCl	7.7	CaMoO$_4$	1.0	GaP	18
CsBr	8.5	PbMoO$_4$	6.1	GaAs	25
CsI	9.9	NH$_4$H$_2$PO$_4$	2.1	InAs	50

[a]For 532 nm: 7.32396 along [010], 7.60968 normal to (100) [65], [b][66]

$$\Lambda_o = \frac{\rho}{\alpha}\left(\frac{\partial\alpha}{\partial\rho}\right) \tag{3.8}$$

Then, the real pressure coefficient is

$$\rho\left(\frac{dn}{d\rho}\right) = (1-\Lambda_o) \times \rho\left(\frac{\partial n}{\partial\rho}\right) \tag{3.9}$$

In most substances, $\rho(\partial n/\partial\rho) > 0$, but in the crystals of diamond, MgO and ZnS this coefficient is negative [29]. In the shock-compressed Al$_2$O$_3$, n increases in the direction of the a-axis, but decreases along the c-axis [31]. The $\rho(dn/d\rho)$ coefficients of alkali halides from [32] are presented in Table 3.9, the dn/dP (in 10^{-2} GPa^{-1}) values from [33–35] for several elements and binary crystalline substances are listed in Table 3.10. The lowest $(1/n)(dn/dP)$ coefficients (the order of 10^{-4} GPa^{-1}) are found in the crystals of c-BN, diamond and SiC, viz. -3.2, -3.4 and -8.3, respectively [36]. The largest increase of RI, from 1.28 at ambient pressure up to 3.31 at $P = 251$ GPa, has been recorded in CsH [37].

For condensed rare gases, hydrogen and water, RIs increase monotonically under high pressures up to 35 GPa [38]. The increase of the RI of H$_2$ under pressures of up to 130 GPa is described by the equation $n = -0.687343 + 0.00407826P + 1.86605$ $(0.29605 + P)^{0.0646222}$ [39]. In BeH$_2$, the RI depends on pressure as $n = 1.474 + 0.0868P - 0.00245P^2$ [40]. In solid methane, RI drastically changes between 208 and 288 GPa, due to an insulator $\rightarrow$ semiconductor phase transition [24]. SiH$_4$ shows two dependences, viz. $n = 1.5089 + 0.00349 \times 10^{-4}P$ from 7 to

Table 3.9 Changes of the RI ($\lambda = \infty$) of alkali halides under compression (10^{-4} GPa^{-1})

MX	LiF	NaF	NaCl	NaBr	NaI	
$\rho(dn/d\rho)$	0.125	0.124	0.276	0.360	0.490	
MX	KCl	KBr	KI	CsCl	CsBr	CsI
$\rho(dn/d\rho)$	0.313	0.353	0.423	0.351	0.416	0.474

Table 3.10 Changes of the RI ($\lambda = \infty$) of crystals under high pressure (10^{-2} GPa^{-1})

Crystal	Si	Ge	NaCl	CsI	MgO	ZnO
$-dn/dP$	0.46	4.5	1.16	3.74	0.158	0.33
Crystal	ZnS	CdS	β-GaS	ϵ-GaSe	γ-InSe	
$-dn/dP$	0.29	1.38	1.18	1.80	1.57	
Crystal	AlN	GaN	GaP	GaAs	InN	
$-dn/dP$	0.17	0.70	1.1	1.3	1.25	

109 GPa and $n = 0.33955 + 0.02332P$ from 109 to 210 GPa, the break at $P = 109$ GPa signifying an insulator $\rightarrow$ semiconductor phase transition [24]. In solid CO_2 under $P \geq 0.6$ GPa, RI increases as $n = 1.41P^{0.041}$ [41].

RI of LiF was studied in great details under high pressures because it is a common window material used in shock-compression experiments. It remained transparent up to 200 GPa [42] and even 800 GPa [43] with the RI increasing to 1.468 and 1.600, respectively, compared to 1.3935 at ambient pressure. It was shown [42] that a modified Gladstone–Dale equation fits the experimental values of RI best.

The results of measurements of RIs in solids under shock compression show that the crystalline state of these substances is preserved at high pressures, but the materials recovered after explosive loading show an enhanced concentration of defects. As a result, the optical anisotropy of the shocked crystals is blurred, even to a quasi-isotropic state [44–50]. Thus, shock-loaded quartz contains amorphous inclusions in the form of planar layers [44]; similarly deformed quartz was found in various meteorite craters. The higher the peak pressure, the more the ellipsoidal optical indicatrix of shocked low-symmetry minerals approximated a sphere, i.e. the crystals becoming quasi-isotropic with a simultaneous drop of the average RI [45, 46]. In the case of pyroxene and sillimanite, the reduction of RIs proved to be due to sub-microscopic inclusions of a strongly disoriented or even amorphous phase, whereas in the shocked quartz, it was attributed to vitrification as an intermediate stage in the quartz-stishovite transformation [47, 50]. The reduction of RI was observed in shock-compressed LnF$_3$ crystals due to the formation of atomic and electronic defects [51].

RIs of various crystalline substances increase with the increase of covalency and metallicity of bonding, since in these cases the absorption bands in the spectra of substances are near to the optical frequency at which the RIs were measured, i.e. $\Delta\omega \rightarrow 0$. Then, it follows from Eq. 1.6 that $n \rightarrow \infty$. In isotropic condensed substances, RIs correlate with their molecular physical properties; thus, the RIs of alkali-halide crystals show a linear dependence on their melting temperatures [52], whereas in organic liquids RIs correlate with boiling temperatures [53].

3.3 Effect of Grain Sizes in Solids on Their Refractive Indices

The study of nanomaterials, i.e. the intermediate state between molecules and crystals, is becoming increasingly important in physics and chemistry, both from the fundamental point of view, and because of their application in different fields of science and technology. Influence of the size factor on the physico-chemical, and especially the molecular physical properties, of inorganic substances is now well researched [54, 55], but optical properties of nanoparticles are as yet insufficiently explored.

From general considerations, it follows that the decrease in the grain sizes from the macro- to micro- to nanorange or a similar reduction in the thickness of a film increases the fraction of surface atoms with lower coordination numbers. In accordance with the experimental data (Table 1.1), this should lead to a decrease of the RI. Indeed, RIs significantly decrease with a reduction of the film thickness or the grain size (Table 3.11).

Rapid developments in advanced photonic devices attracted interest to high-RI polymers (HRIP), where high RIs have been achieved either by introducing substituents with high molar refractions (intrinsic HRIPs) or by combining high-RI nanoparticles with polymer matrices, such as polyimides, methacrylates, halogen-,

Table 3.11 Size effect in refractive indices of solids

Solids	$D_{initial}$, nm	D_{final}, nm	$RI_{initial}$	RI_{final}	Refs.
$CdFe_2O_4$	1031	346	2.15	1.88	[67]
Si	bulk	36	3.44	1.76	[68]
CuO	310	125	2.80	2.70	[69]
ZnO	120	20	1.96	1.58	[70]
ZnSe	677	275	3.43	2.93	[71]
ZnTe	695	331	3.37	3.13	[72]
CdS	21.5	16.3	2.06	1.71	[73]
CdSe	bulk	3.5	2.65	2.34	[74]
SnS	585	155	2.55	2.17	[75, 76]
NiO	270	60	2.02	1.97	[77]
CdI_2	696	272	2.145	2.100	[78]
ZrO_2	120	17	2.2	2.02	[79, 80]
MoO_x^a	600	35	1.73	2.08	[81]

[a]This exception was explained by the Wemple–DiDomenico model (see above)

sulphur-, phosphorus- and silicon-containing materials (HRIP nanocomposites). The Lorentz–Lorenz equation (3.4) is often applied to predict RI of a polymer using the molar refraction (R), molecular mass (M) and molar volume (V) of the polymer. From Eq. 3.4 it follows that

$$n = \sqrt{\frac{1 + 2R/V}{1 - R/V}} \tag{3.10}$$

Thus, according to Eq. 3.10, introducing substituents with high molar refractions and low molar volumes can efficiently increase the RI of a polymer. The most commonly used nanoparticles for HRIPs include TiO_2 (anatase, $n = 2.450$; rutile, $n = 2.571$), ZrO_2 ($n = 2.20$), amorphous silicon ($n = 4.23$), PbS ($n = 4.20$), and ZnS ($n = 2.36$). In order to achieve good optical transparency and avoid Rayleigh scattering of the nanocomposite, the diameter of the nanoparticle should be below 25 nm [56].

Thus, ZnS nanoparticles with a diameter of 2–6 nm were dispersed in the polythiourethane (PTU) matrix, and the resulting nanocomposite films showed the RI (at 633 nm) ranging from 1.574 to 1.848 and increasing linearly with the ZnS content [57]. High-RI polyimide–titania hybrid optical thin films were prepared with a homogeneous structure and nanoscale size of the TiO_2 particles, and excellent optical transparency in the visible region was obtained [58]. The RI at 633 nm of such hybrid thin films increases linearly from 1.66 to 1.82 with increasing TiO_2 content from 0 to 40 vol.%. Recently, graphene has been used as nanoparticles in nanocomposite HRIPs, resulting in a promising RI of 2.058 [59]. One can expect that the upper limit of intrinsic polymer RIs has yet to be reached. In particular, phosphorus, silicon, fullerenes and organometallic components remain understudied, as well as the interfacial areas combining two or more of these high molar refractivities [60].

References

1. I. Newton, *Opticks, London B.* **2**, 204 (1704)
2. M. Beer, *Einleitung in hohere Optik, Brunswick* 35 (1853)
3. T. Gladstone, T. Dale, Philos. Trans. **153**, 317 (1863)
4. L. Lorenz, Kgl. Dansk. Videns. Selsk. Skrift. **10**, 485 (1875)
5. L. Lorenz, Wied. Ann. **11**, 70 (1880)
6. H.A. Lorentz, Verhl. Akad. Wetens, Amsterdam **18**, 60 (1879)
7. H.A. Lorentz, Wied. Ann. **9**, 641 (1880)
8. S.S. Batsanov, *Refractometry and chemical structure* (Van Nostrand, Princeton, 1966)
9. S.S. Batsanov, *Structural refractometry*, 2nd edn. (Vysshaya Shkola, Moscow, 1976). (in Russian)
10. S.S. Batsanov, A.S. Batsanov, *Introduction to structural chemistry* (Springer, Dordrecht, 2012)
11. M.M. Yakshin, Ann. Inst. Platine (USSR) **21**, 146 (1948)

12. I. Chernyaev, Ann. Inst. Platine (USSR) **4**, 243 (1926)
13. I. Chernyaev, Proc. Acad. Sci. USSR Chem. Div. **2**, 197 (1953)
14. G.B. Kauffmann, J. Chem. Educ. **54**, 86 (1977)
15. B.J. Coe, S.J. Glenwright, Coord. Chem. Rev. **203**, 5 (2000)
16. G.B. Bokii, S.S. Batsanov, Doklady Acad. Sci. USSR **95**, 1205 (1954). (in Russian)
17. S.S. Batsanov, Rus. J. Inorg. Chem. **2**, 2018 (1957). (in Russian)
18. S.S. Batsanov, E.D. Ruchkin, *ibid*, **2**, 2553 (1957). (in Russian)
19. S.S. Batsanov, O.P. Alexandrova, *ibid*, **3**, 2666 (1958). (in Russian)
20. B. Pinter, V. Van Speybroeck, M. Waroquier, P. Geerlings, F. De Proft, Phys. Chem. Chem. Phys. **15**, 17354 (2013)
21. F. Guégan, V. Tognetti, L. Joubert, H. Chermette, D. Luneau, C. Morell, Phys. Chem. Chem. Phys. **18**, 982 (2016)
22. D. Goldhammer, *Dispersion und Absorption des Lichtes* (Tubner-Ferlag, Leipzig, 1913)
23. K. Herzfeld, K. Phys. Rev. **29**, 701 (1927)
24. L. Sun, A.L. Ruoff, C.-S. Zha, G. Stupian, J. Phys. Chem. Solids **67**, 2603 (2006)
25. S.S. Batsanov, Rus. J. Inorg. Chem. **49**, 560 (2004)
26. S.D. Hamann, M. Linton, Trans. Faraday Soc. **62**, 2234 (1966)
27. S.S. Batsanov, J. Engin. Phys. **12**, 59 (1967)
28. R.M. Waxler, G.W. Cleek, J. Res. NBS **A77**, 755 (1973)
29. M. Bass (ed.), *Handbook of optics*, vol. 2, 2nd edn. (McGraw-Hill, New York, 1995)
30. H. Mueller, Phys. Rev. **47**, 947 (1935)
31. S.C. Jones, M.C. Robinson, Y.M. Gupta, J. Appl. Phys. **93**, 1023 (2003)
32. P.G. Johannsen, G. Reiss, U. Bohle et al., Phys. Rev. B **55**, 6865 (1997)
33. A.R. Goñi, F. Kaess, J.S. Reparaz et al., Phys. Rev. B **90**, 045208 (2014)
34. R. Oliva, A. Segura, J. Ibáñez et al., Appl. Phys. Lett. **105**, 232111 (2014)
35. F.J. Manjón, Y. van der Vijver, A. Segura, V. Muñoz, Semicond. Sci. Technol. **15**, 806 (2000)
36. N.M. Balzaretti, J.A.H. da Jornada, Solid State Commun. **99**, 943 (1996)
37. K. Ghandehari, H. Luo, A.L. Ruoff et al., Solid State Commun. **95**, 385 (1995)
38. A. Dewaele, J.H. Eggert, P. Loubeyre, R. LeToullec, Phys. Rev. B **67**, 094112 (2003)
39. W.J. Evans, I.J. Silvera, Phys. Rev. B **57**, 14105 (1998)
40. M. Ahart, J.L. Yarger, K.M. Lantzky et al., J. Chem. Phys. **124**, 014502 (2006)
41. H. Shimizu, T. Kitagawa, S. Sasaki, Phys. Rev. B **47**, 11567 (1993)
42. P.A. Rigg, M.D. Knudson, R.J. Scharff, R.S. Hixson, J. Appl. Phys. **116**, 033515 (2014)
43. D.E. Fratanduono, T.R. Boehly, M.A. Barrios et al., J. Appl. Phys. **109**, 123521 (2011)
44. W. Engelhardt, W. Bertsch, Contrib. Miner. Petrol. **20**, 203 (1969)
45. D. Stöffler, Fortschr. Miner. **49**, 50 (1972)
46. D. Stöffler, Fortschr. Miner. **51**, 256 (1974)
47. H. Schneider, U. Hornemann, Contrib. Mineral. Petrol. **55**, 205 (1976)
48. R. Jeanloz, T.J. Ahrens, J.S. Lally et al., Science **197**, 457 (1977)
49. M. Jakubith, G. Lehmann, Ber. Bunsen. Phys. Chem. **83**, 609 (1979)
50. J.R. Ashworth, H. Schneider, Phys. Chem. Miner. **11**, 241 (1985)
51. S.S. Batsanov, E.V. Dulepov, E.M. Moroz et al., Comb. Expl. Shock Wave **7**, 226 (1971)
52. S.S. Batsanov, Crystallography **1**, 140 (1956). (in Russian)
53. M.M. Samygin, Rus. J. Phys. Chem. **11**, 325 (1938). (in Russian)
54. E. Rodumer, Chem. Soc. Rev. **35**, 583 (2006)
55. S.S. Batsanov, J. Struct. Chem. **52**, 602 (2011)
56. J.-G. Liu, M. Ueda, J. Mater. Chem. **19**, 8907 (2009)
57. C. Lu, Z. Cui, Z. Li et al., J. Mater. Chem. **13**, 526 (2003)
58. C.-M. Chang, C.-L. Chang, C.-C. Chang, Macromol. Mater. Eng. **291**, 1521 (2006)
59. G. Zhang, H. Zhang, X. Zhang et al., J. Mater. Chem. **22**, 21218 (2012)
60. E.K. Macdonald, M.P. Shaver, Polym. Int. **64**, 6 (2015)
61. A.I. Vogel, *J. Chem. Soc* 1833 (1948)
62. A.I. Vogel, W.T. Cresswell, G. Jeffery, J. Leicester, J. Chem. Soc. 514 (1952)
63. K.J. Miller, J. Am. Chem. Soc. **112**, 8533 (1990)

64. S.S. Batsanov, G.N. Kustova, E.D. Ruchkin, V.S. Grigorieva, J. Struct. Chem. **6**, 47 (1965)
65. I. Bhaumik, R. Bhatt, S. Ganesamoorthy et al., Appl. Opt. **50**, 6006 (2011)
66. Y.I. Vesnin, S.S. Batsanov, J. Struct. Chem. **6**, 501 (1965)
67. E.R. Shaaban, Appl. Phys. A **115**, 919 (2014)
68. D. Amans, S. Callard, A. Gagnaire et al., J. Appl. Phys. **93**, 4173 (2003)
69. Y. Akaltun, Thin Solid Films **594**, 30 (2015)
70. M. Gilliot, A. Hadjadj, J. Martin, Thin Solid Films **597**, 65 (2015)
71. E.R. Shaaban, J. Alloys Compd. **563**, 274 (2013)
72. E.R. Shaaban, I. Kansal, S.H. Mohamed, J.M.F. Ferreira, Physica B **404**, 3571 (2009)
73. S. Kumar, S. Kumar, P. Sharma et al., J. Appl. Phys. **112**, 123512 (2012)
74. C. Gan, Y. Zhang, S.W. Liu et al., Opt. Mater. **30**, 1440 (2008)
75. A. Jakhar, A. Jamdagni, A. Bakshi et al., Solid State Commun. **168**, 31 (2013)
76. M.S. Selim, M.E. Gouda, M.G. El-Shaarawy et al., Thin Solid Films **527**, 164 (2013)
77. E.R. Shaaban, M.A. Kaid, M.G.S. Ali, J. Alloys Compd. **613**, 324 (2014)
78. I.S. Yahi, M. Shapaan, Y.A.M. Ismail et al., J. Alloys Compd. **636**, 317 (2015)
79. S. Zhao, F. Ma, Zh. Song, K. Xu, Opt. Mater. **30**, 910 (2008)
80. X. Wang, G. Wu, B. Zhou, J. Shen, J. Alloys Compd. **556**, 182 (2013)
81. Ü. Akın, H. Şafak, J. Alloys Compd. **647**, 146 (2015)

Part II
Anhydrous Substances

Chapter 4
Refractive Indices of Elements and Binary Compounds

Tables 4.1, 4.2, 4.3, 4.4, 4.5, 4.6 and 4.7 list the RIs of elements, such binary compounds as hydrides, halides, oxides, chalcogenides and pnictides of metals, as well as hydroxides and cyanide salts. For non-metallic elements and for most of binary compounds, RI was measured at 589.3 nm (n_D), except where stated otherwise. For metallic elements, and for compounds where the dispersion of the RI was studied, the results are extrapolated to $\lambda = \infty$ and are given as n_∞. The data quoted from Refs. [1–5] and the authors' own (previously unpublished) measurements are presented without references, and in other cases, it will be cited sources of information.

Table 4.1 Refractive indices in metals [4]

M	n_{max}	M	n_{max}	M	n_{max}
Be[a]	9.68	Re	4.25	Cu	29.7
Mg[b]	30.3	Fe	6.41	Ag[a]	16.3
Ti[c]	36.2	Co	6.71	Au[a]	15.3
V[d]	2.77	Ni	9.54	Zn[e]	15.3
Nb	16.0	Ru	11.7	Hg	14.0
Ta[c]	56.4	Rh	18.5	B	3.08
Cr	21.2	Pd	4.13	Al[a]	24.2
Mo	18.5	Os	4.08	In[f]	24.8
W	14.1	Ir	28.5	Sn	4.8
Mn[d]	3.89	Pt	13.2	Pb	13.6

[a][6], [b][7], [c][8], [d][9], [e][10] and [f][11]

© The Author(s) 2016
S.S. Batsanov et al., *Refractive Indices of Solids*,
SpringerBriefs in Applied Sciences and Technology,
DOI 10.1007/978-981-10-0797-2_4

Table 4.2 Refractive indices of nonmetals [4]

M	n_g	n_m	n_p	M	n_g	n_m	n_p
C[a]		2.417		S[j]	2.24	2.04	1.96
C[b]		2.15	1.81	S[k]		2.06	
Si[c]		3.453		Se[h]	2.91	2.84	
Ge[c]		4.104		Se[l]	3.547	2.744	
P[e]		2.12		Te[l]	6.372	4.929	
P[f]	3.15	2.72		I[m]		3.34	
P[g]	3.20	2.72		He[n]		0.897	
P[h]	3.21	3.20	3.11	Ne[n]		0.986	
As		3.6		Ar[o]		1.291	
Sb		10.4		Kr[p]		1.367	
S[i]		2.02		Xe[q]		1.485	
				p-H$_2$[r]		1.130	

[a]Diamond, n_D; [b]graphite, [c]diamond structure, $\lambda = 2$ μm [12]; [d]for $\lambda = 10$ μm; phases: [e]cubic, [f]tetragonal, [g]hexagonal, [h]triclinic, [i]glass, [j]orthorhombic, [k]monoclinic, [l]trigonal, $\lambda = 4$ μm [5]; [m]at $\lambda = 1.8$ μm, liquid iodine (114 °C) $n = 1.934$; liquid bromine (19 °C) $n = 1.604$; [n]extrapolated to normal pressure [13], [o]at $T = 20$ K, [p]at $T = 67$ K, [q]at $T = 80$ K [14] and [r]at $T = 4.4$ K and $\lambda = 514$ nm [15]

Table 4.3 Refractive indices of the MX[a] type compounds

M	F		Cl		Br		I	
	n_D	n_∞	n_D	n_∞	n_D	n_∞	n_D	n_∞
Li	1.392	1.386	1.662	1.646	1.784	1.752	1.955	1.906
Na	1.326	1.320	1.544	1.528	1.641	1.613	1.774	1.730
K	1.362	1.355	1.490	1.475	1.559	1.537	1.667	1.628
Rb	1.396	1.389	1.494	1.472	1.553	1.523	1.647	1.605
Cs[b]	1.478	1.469	1.534	1.517	1.582	1.558	1.661	1.622
Cs[c]	1.578	1.566	1.642	1.619	1.698	1.669	1.788	1.743
NH$_4$	1.315	1.312	1.643	1.614	1.712	1.672	1.701	1.633
Cu			1.973	1.882	2.116	1.974	2.345	2.191
Ag	1.80	1.73	2.067	2.002	2.258	2.166	2.216[d]	2.13
Tl	2.055		2.247	2.162	2.418	2.302	2.78	2.60
M	O		S		Se		Te	
Be	1.724	1.679	2.275				2.65	
Mg	1.737	1.718	2.271	2.084	2.42			
Ca	1.837	1.804	2.137	2.020	2.274	2.148	2.51	
Sr	1.870	1.802	2.107	1.927	2.220	2.092	2.41	
Ba	1.980	1.883	2.155	2.075	2.268	2.146	2.44	
Zn	2.018	1.922	2.368	2.267	2.611	2.429	3.060	2.698
Cd	2.37	2.15	2.514	2.31	2.650	2.454	2.982	2.683

(continued)

Table 4.3 (continued)

M	F		Cl		Br $^\bullet$		I	
	n_D	n_∞	n_D	n_∞	n_D	n_∞	n_D	n_∞
Hg	2.50^e		3.058^f	2.512	3.460		3.753	
Cu	2.634	2.54	2.44$^\alpha$		2.75$^\gamma$	2.47$^\gamma$		
Ga			2.715^g	2.26	2.94^h			
Eu	2.35		2.43	2.20	2.51	2.29	2.70	2.42
Ge	2.62$^\beta$		3.480	3.267^i	4.25		6.00	
Sn	2.78		3.61		3.80	3.680^j	6.32	
Pb	2.621^k		4.280	4.10	4.624^l	4.54	6.038^m	5.73
Mn	2.18		2.67^n		3.12	2.83	3.26^o	
Fe	2.32							
Co	2.30							
Ni	2.27^p		2.325^q					

M	N		P		As		Sb	
B	2.117^r		3.100	2.78				
Al	2.154	2.043	2.75		2.995^s	2.856	4.030^t	
Ga	2.398^u	2.312	3.361	3.08	3.419^v	3.30	4.115^w	3.74
In	3.12^x	2.90	3.587	3.10	3.89	3.76^y	4.425$^\alpha$	3.96^y

$^a n_\infty$ [16]; ^{b}for $N_c = 6$; ^{c}for $N_c = 8$; ^{d}for cubic phase, for hexagonal $n_o = 2.218$, $n_e = 2.229$; ^{e}average of $n_g = 2.65$, $n_m = 2.50$, $n_p = 2.37$; ^{f}average of $n_o = 2.941$, $n_e = 3.307$; g) $\lambda = 633$ nm [5]; ^{h}[17]; ^{i}average of $n_g = 3.358$ and $n_p = 3.176$ for $\lambda = \infty$ [18]; ^{j}average of $n_o = 3.882$, $n_e = 3.308$ [19]; ^{k}average of $n_m = 2.665$, $n_p = 2.535$ for$\lambda = 671$ nm; $^l\lambda = 1.06$ µm; $^m\lambda = 4$ µm; ^{n}for cubic phase, the average for hexagonal phase is $n = 2.45$; ^{o}[20], ^{p}at $\lambda = 671$ nm, ^{q}average of $n_g = 3.22$, $n_m = 2.046$, $n_p = 1.908$; ^{r}for cubic phase; for hexagonal phase of the graphite type $n_o = 2.08$, $n_e = 1.72$; ^{s}at $\lambda = 1.4$ µm; ^{t}[21]; ^{u}for cubic phase, for w-GaN: $n_o = 2.44$, $n_e = 2.40$ [22]; ^{v}at $\lambda = 2.5$ µm; ^{w}at $\lambda = 0.99$ µm; ^{x}at $\lambda = 0.66$ µm; $^y n_\infty$ [16]; ^{z}at $\lambda = 825$ nm; $^\alpha$[23], $^\beta$[24] and $^\gamma$at $\lambda = 0.70$ µm [25]

Additions

LiH $n_D = 1.615$, NaH $n_D = 1.470$, KH $n_D = 1.453$; CsH $n_D = 1.28$ [26]; LiOH $n_o = 1.464$, $n_e = 1.452$; NaOH $n_g = 1.472$, $n_m = 1.470$, $n_p = 1.457$; KOH $n_g = 1.497$, $n_m = 1.492$, $n_p = 1.486$; NaCN $n_D = 1.452$, KCN $n_D = 1.410$, CuCN $n_g = 2.07$, $n_m = 1.80$, $n_p = 1.73$; AgCN $n_g = 1.94$, $n_p = 1.685$; TlCN $n = 2.02$ [27, 28]; NaSCN $n_g = 1.695$, $n_m = 1.625$, $n_p = 1.545$; KSCN $n_g = 1.730$, $n_m = 1.660$, $n_p = 1.532$; ScN $n = 2.46$ [29]; SiO $n_D = 2.04$ [30]; α-SiC: $n_g = 2.691$, $n_m = 2.648$; β-SiC $n_D = 2.647$, $n_\infty = 2.553$ [5]; NiAs $n_o = 2.11$, $n_e = 1.80$ [31]

Table 4.4 Refractive indices of the MX_2 type solids (o = orthorhombic, tet = tetragonal)

Compounds	n_g	n_m	n_p	Compounds	n_g	n_m	n_p
BeH_2		1.648		TlClS		2.18	
BeF_2^a		1.275		TlBrS		2.46	
BeF_2^b		1.328		TlIS		2.7	
BeF_2^c		1.345		TlClSe		2.30	
BeI_2^d		1.99		TlBrSe^t		2.51	
BeI_2^e	1.988	1.954	1.952	TlSeBr^u		2.34	
$Be(OH)_2$	1.548	1.544	1.539	TlISe		>2.7	
MgH_2	1.96	1.95		tet-YOF		1.728	1.682
MgF_2	1.389	1.377		o-YOF		1.76	
$MgCl_2$		1.675	1.590	YO(OH)		1.845	
$Mg(OH)_2$	1.585	1.566		LaO(OH)		1.798	
CaF_2		1.434		NdOF		1.82	
CaFCl		1.668	1.635	NdO(OH)		1.85	
$CaCl_2$	1.613	1.605	1.600	tet-SmOF		1.78	1.74
$Ca(OH)_2$	1.577	1.550		SmOF		1.82	
CaI_2		1.743	1.652	SmO(OH)	1.924	1.855	
SrF_2^f		1.438^f		GdOF		1.80	
SrF_2^g		1.482		TbOF		1.77	
SrFCl		1.651	1.627	tet-DyOF		1.760	1.721
$SrCl_2$		1.691		o-DyOF		1.81	
$Sr(OH)_2$	1.610	1.599	1.588	tet-HoOF		1.757	1.718
BaF_2^f		1.474		o-HoOF		1.80	
BaF_2^g		1.518		tet-ErOF		1.750	1.718
BaFCl		1.640	1.633	o-ErOF		1.79	
$BaCl_2$	1.742	1.736	1.730	tet-YbOF		1.747	1.716
$BaBr_2$		1.793		o-YbOF		1.78	
BaO_2	1.85	1.775		LaSF	>2.14	2.06	
ZnF_2	1.525	1.495		CeSF	>2.14	2.03	
ZnFCl		1.70		PrSF	>2.14	2.10	
$ZnCl_2$	1.713	1.687		NdSF	>2.14	2.04	
$ZnBr_2$	1.842	1.825		CO_2^v		1.412	
$Zn(OH)_2$	1.538	1.532	1.526	SiO_2^a		1.459	
CdF_2		1.562		SiO_2^w	1.473	1.469	1.468
$CdCl_2$		1.850	1.714	SiO_2^x		1.484	
$CdBr_2$		2.027	1.866	SiO_2^y		1.487	1.484
CdI_2		2.36	2.17	SiO_2^z	1.522	1.513	
$Cd(OH)_2$		1.802	1.70	SiO_2^a	1.540	1.533	
$CdAs_2^i$	3.916	3.555		SiO_2^β	1.553	1.544	
$HgCl_2$	1.965	1.859	1.725	SiO_2^c	1.599	1.595	
$HgBr_2$	2.095	1.922	1.879	SiO_2^p	1.835	1.800	
HgI_2		2.748	2.455	GeO_2^a		1.607	

(continued)

Table 4.4 (continued)

Compounds	n_g	n_m	n_p	Compounds	n_g	n_m	n_p
$Hg_2Cl_2^i$	2.656	1.974		GeO_2^b		1.735	1.695
Hg_2OCl	2.66	2.64	2.35	GeO_2^p		2.07	1.99
Hg_2OCl_2		2.21	2.19	GeS_2^γ		2.30	2.25
$Hg(CN)_2$		1.645	1.492	$GeSe_2^\gamma$	3.32	2.83	2.65
CuF_2		1.527	1.515	SnO_2	2.097	2.001	
$Cu(OH)_2^k$	>2.03	?	1.720	$SnOS$		2.67	
SmF_2		1.636		SnS_2		2.85	2.16
EuF_2		1.555		$SnSe_2^\delta$		3.26	2.88
YbF_2		1.618		PbO_2		2.23	
$GaO(OH)^l$		1.827		TiO_2^ϵ		2.561	1.488
SnF_2^m	1.878	1.831	1.800	TiO_2^η	2.700	2.584	2.583
PbF_2^n		1.767		TiO_2^i	2.908	2.621	
PbF_2^g	1.853	1.844	1.837	TiS_2^θ		3.70	
$PbFCl$		2.145	2.006	$TiSe_2^\theta$		5.92	
$PbCl_2$	2.260	2.217	2.199	ZrO_2	2.20	2.19	2.13
$PbOHCl$	2.158	2.116	2.077	ZrS_2		3.14	1.74
$PbBr_2$	2.560	2.476	2.439	$ZrTe_2$		3.13	
PbI_2		2.80	2.13	HfO_2		2.122	
$Pb(N_3)_2$	2.64	2.24	1.86	HfS_2		2.49	
CrF_2	1.547	1.520		$HfSe_2$		2.84	
$CrO(OH)^o$		2.155	1.975	CeO_2		2.42	
MnF_2^g	1.507	1.480		ThO_2^i		2.170	
MnF_2^s	1.492	1.490	1.484	UO_2 •		2.39	
$Mn(OH)_2$		1.723	1.681	PuO_2		2.402	
$MnO(OH)$	2.53	2.25	2.25	PbO_2		2.23	
$MnCl_2$		1.80		VO_2^κ		2.59	
FeF_2	1.528	1.518		TeO_2^s	2.430	2.274	
$FeO(OH)^A$	2.409	2.403	2.274	MnO_2		2.25	
$FeO(OH)^B$	2.515	2.200	1.938	MoS_2^h		4.336	2.035
$Fe(OH)_2^q$		1.722	1.707	$MoSe_2$		4.22	
$FeCl_2$		1.567		WSe_2		4.5	
CoF_2	1.547	1.520		MnS_2^h		2.634	
NiF_2	1.562	1.530		FeS_2		3.07	
$Ni(OH)_2^r$		1.760	1.759	H_2O^λ	1.311	1.309	
$PdCl_2^h$		2.17	2.14_5	Li_2O^μ		1.644	
$PdCl_2^h$	2.50	2.04	1.75	Na_2O^μ		1.532	
$PtCl_2$ ^{h}		2.14	2.05_5	Cu_2O^π		3.041	
$PtBr_2^h$	2.55	2.47	2.43	Cu_2S		3.52	3.49
$Pt(SCN)_2$		1.93		Ag_2S		3.55	
Be_2C		2.635		$HfSi_2$ ^{s}		3.249	
$AlON$ ^{s}		1.789		$ZrSi_2$ ^{s}		3.544	

(continued)

Table 4.4 (continued)

Compounds	n_g	n_m	n_p	Compounds	n_g	n_m	n_p
AlO(OH)	1.750	1.722	1.702	VSi$_2^s$		3.849	
SiON ^{s}		1.483		TaSi$_2^s$		3.607	
Mg$_2$Si		3.65		MoSi$_2^s$		4.752	4.732
Mg$_2$Ge		3.73		ReSi$_2^s$		4.504	
Mg$_2$Sn		3.94		FeSi$_2^s$		3.991	
				CoSi$_2$ ^{s}		3.096	

[a]Glass, [b]type of quartz, [c]type of coesite, [d]tetragonal phase, [e]orthorhombic phase, [f]structure of CaF$_2$, [g]structure of PbCl$_2$, [h]at λ = 852 nm; α-Pb(N$_3$)$_2$ n_g = 2.64, n_m = 2.24, n_p = 1.86; β-Pb(N$_3$)$_2$ n_g = 2.70, n_m = 2.14, n_p = 1.98 [32], [i][33], [j][34], [k][35], [l][36], [m]at $\lambda = \infty$ n_g = 1.8105, n_m = 1.7749, n_p = 1.7505; [n]cubic phase, [o][37], [p]type of rutile, [q][38]; [r][39], [s][5], [t]for Se = Tl^{III}– Br, [u]for Tl^I–Se–Br, [v][40], [w]trydimite, [x]β-crystabalite, [y]α-crystobalite, [z]ceatite, [α]β-quartz, [β]α-quartz, [γ][3], [δ]for glass, n_∞ = 3.142 [41], [ε]anatase, [η]broocite, [θ][42], [ι]at λ = D, n_∞ = 2.119; [κ][43], [λ][5], [μ][44], [π]n_∞ = 2.557, [A]goetite and [B]lepidocrite

Table 4.5　Refractive indices of the MX$_3$ type solids

Compounds	n_g	n_m	n_p	Compounds	n_g	n_m	n_p
ScF$_3$		1.401		BiF$_3$		1.86	
YF$_3^a$	1.570	1.550	1.536	CrF$_3$	1.582	1.568	
LaF$_3^a$		1.603	1.597	CrCl$_3$		1.64	
CeF$_3$		1.618	1.612	FeF$_3$	1.552	1.541	
PrF$_3$		1.618	1.614	CoF$_3$	1.726	1.703	
NdF$_3^a$		1.621	1.617	RhF$_3$		1.92$_5$	
SmF$_3$	1.607	1.594	1.575	UF$_3$	1.738	1.732	
EuF$_3$		1.590		UCl$_3$	2.08	1.965	
GdF$_3^a$	1.601	1.586	1.567	PuF$_3$	1.685	1.684	
TbF$_3$	1.602	1.588	1.569	B(OH)$_3$	1.462	1.461	1.337
DyF$_3^a$	1.601	1.587	1.568	Al(OH)$_3$	1.587	1.566	1.566
HoF$_3^a$	1.599	1.585	1.565	AlF(OH)$_2$	1.567	1.552	1.532
ErF$_3^a$	1.598	1.581	1.563	Ga(OH)$_3^b$		1.736	
TmF$_3$	1.597	1.578	1.562	Y(OH)$_3$	1.714	1.676	
YbF$_3^a$	1.595	1.575	1.558	La(OH)$_3$		1.768	1.740
LuF$_3$	1.588	1.569	1.553	Nd(OH)$_3$		1.800	1.755
UF$_3$	1.738	1.732		Sm(OH)$_3$		1.800	1.758
AlF$_3$	1.377	1.376		Eu(OH)$_3$		1.735	
GaF$_3$		1.438		In(OH)$_3$		1.716	
InF$_3$		1.449	1.430	Fe(OH)$_3^c$	1.94	?	1.92
BBr$_3$		1.531		VO(OH)$_2^d$	2.01	1.900	1.810
BI$_3^b$		2.32		MoO$_3$	2.37	2.27	2.25
AsI$_3$		2.59	2.23	WO$_3$	2.703	2.376	2.283
SbF$_3$	1.667	1.620	1.574	NLi$_3^e$		2.12	

(continued)

Table 4.5 (continued)

Compounds	n_g	n_m	n_p	Compounds	n_g	n_m	n_p
SbI_3		2.78	2.36	PLi_3^e		2.19	
C_2Cl_6	1.668	1.598	1.590	$AsLi_3^e$		2.28	
VF_3	1.544	1.536		AsI_3	2.59	2.23	

[a][45], [b][46], [c][47], [d][48] and [e][49]

Table 4.6 Refractive indices of M_2X_3 type solids

M_2O_3	n_m	n_p	M_2X_3	n_g	n_m	n_p
$B_2O_3^a$	1.447		$B-Tm_2O_3$		2.015	
$B_2O_3^b$	1.468		$C-Tm_2O_3$		1.951	
$B_2O_3^c$	1.458		$B-Yb_2O_3$		2.00	
$B_2O_3^d$	1.648		$C-Yb_2O_3$		1.947	
$Al_2O_3^d$	1.769	1.760	$B-Lu_2O_3$		1.99	
Al_2O_3 [d]	1.67	1.64	$C-Lu_2O_3$		1.927	
$Al_2O_3^c$	1.696		$As_2O_3^e$	2.01	1.92	1.87
$Ga_2O_3^c$	1.927		$As_2O_3^c$		1.755	
$In_2O_3^c$	2.093		$Sb_2O_3^c$		2.087	
Sc_2O_3	1.994		$Sb_2O_3^f$	2.36	2.35	2.18
$B-Y_2O_3$	1.97		$Bi_2O_3^e$		2.45	
$C-Y_2O_3$	1.915		$Bi_2O_3^c$		2.42	
La_2O_3	2.03		Cr_2O_3 [d]		2.49	2.47
Pr_2O_3	1.94		Mn_2O_3		2.33	
Nd_2O_3	2.00		Fe_2O_3 [d]		3.22	2.94
$B-Sm_2O_3$	2.08		Y_2S_3		2.61	
$B-Eu_2O_3$	2.07		La_2S_3		2.85	
$C-Eu_2O_3$	1.983		Ho_2S_3		2.63	
$B-Gd_2O_3$	2.04		Yb_2S_3		2.61	
$C-Gd_2O_3$	1.977		$As_2S_3^g$	3.02	2.81	2.40
Tb_2O_3	1.964		$As_2S_3^a$		2.648	
$B-Dy_2O_3$	2.035		$As_2Se_3^a$		3.3	
$C-Dy_2O_3$	1.974		$Sb_2S_3^f$	4.303	4.046	3.194
$B-Ho_2O_3$	2.03		$Sb_2Se_3^g$		3.20	
$C-Ho_2O_3$	1.963		$Sb_2Te_3^g$		9.0	
$B-Er_2O_3$	2.025					

[a]Glass, [b]dense glass, [c]cubic phase, [d]hexagonal phase, [e]orthorhombic phase, [f]monoclinic phase and [g]trigonal phase

Table 4.7 Refractive indices of the M_mX_n type solids

MX_4	Type	n_g	n_m	n_p	MX_4	Type	n_g	n_m	n_p
CeF_4	ZrF_4	1.652	1.613	1.607	PCl_5	tetragonal	1.708	1.674	
	LaF_3		1.632	1.629	$PtCl_2Br_2$	–		2.07	
ZrF_4	ZrF_4	1.60	1.57		$PtBr_2I_2$	–		2.09	2.01
HfF_4	ZrF_4		1.58	1.54	$Pb_3O_4^c$			2.40	
ThF_4	ZrF_4		1.53		V_2O_5		2.89	2.55	2.10
	LaF_3		1.613	1.610	$Nb_2O_5^e$			2.340	
UF_4	cubic		1.685		$Ta_2O_5^e$			2.146	
	ZrF_4	1.594	1.584	1.550	$P_2O_5^d$	hexagonal	1.471	1.469	
UF_3Cl	monocl.	1.755	1.745	1.725	$P_2O_5^d$	orthorhom.	1.589	1.578	1.545
UCl_4	$ThCl_4$		2.03	1.92	$P_2O_5^d$	tetragonal	1.624	1.599	
UBr_4	$ThCl_4$	2.06	2.02	1.86	Fe_3O_4			2.42	
PuF_4	ZrF_4	1.629	1.612	1.577	Al_2OC^f	hexagonal	2.16	2.11	
CCl_4^a	$SnCl_4$		1.519		$Al_4O_4C^f$	orthorhom.	2.010	?	1.985
$GeBr_4$	$SnCl_4$		1.627		Al_4C_3		2.75	2.70	
SnI_4	cubic		2.106		$Si_3N_4^e$			2.017	
$SeCl_4$	cubic		1.807		Pt $(CH_3)_3Cl^f$			1.74	
$NbCl_5^b$	–	2.08	2.02	2.01	B_4C^g			3.19	

[a]T = 243 K [50], [b][51], [c]at λ = 671 nm, [d][52], [e][53], [f][54], [e][55], [f][56] and [g]at λ = 18.5 μm [57]

References

1. E. Kordes, *Optische Daten* (Verlag Chemie, Weinheim, 1960)
2. A.N. Winchell, H. Winchell, *Optical properties of artifical minerals* (Academic Press, New York, 1964)
3. S.S. Batsanov, A.S. Batsanov, *Introduction to structural chemistry* (Springer, Dordrecht, 2012)
4. D.R. Lide (ed.), *Handbook of chemistry and physics*, 88th edn. (CRC Press, Boca Raton, 2007–2008)
5. M. Bass (ed.), *Handbook of optics,*, vol. 2, 2nd edn. (McGraw-Hill, New York, 1995)
6. A.D. Rakić, A.B. Djurišić, J.M. Elazar, M.L. Majewski, Appl. Opt. **37**, 5271 (1998)
7. H.-J. Hagemann, W. Gudat, C. Kunz, J. Opt. Soc. Amer. **65**, 742 (1975)
8. M.A. Ordal, R.J. Bell, R.W. Alexander et al., Appl. Opt. **27**, 1203 (1988)
9. P.B. Johnson, R.W. Christy, Phys. Rev. B **9**, 5056 (1974)
10. G.P. Motulevich, A.A. Shubin, Sov. Phys. JETP **29**, 24 (1969)
11. A.I. Golovashkin, I.S. Levchenko, G.P. Motulevich, A.A. Shubin, Sov. Phys. JETP **24**, 1093 (1967)
12. A. Dewaele, J.H. Eggert, P. Loubeure, R. Le Toullec, Phys. Rev. B **67**, 094112 (2003)
13. A.C. Sinnoc, J. Phys. **C13**, 2375 (1980)
14. N.M. Balzaretti, J.A.H. Da Jordana, J. Phys. Chem. Solids **57**, 179 (1996)
15. M. Perera, B.A. Tom, Y. Miyamoto et al., Opt. Letters **36**, 840 (2011)
16. M. Gauthier, A. Polian, J. Besson, A. Chevy, Phys. Rev. B **40**, 3837 (1989)
17. A. Elkorashy, Phys. B **159**, 171 (1989)
18. A. Elkorashy, J. Phys. Chem. Solids **51**, 289 (1990)
19. S. Zollner, Ch. Lin, E. Schönherr et al., J. Appl. Phys. **66**, 383 (1989)

20. G.J. Glanner, H. Sitter, W. Faschinger, M.A. Herman, Appl. Phys. Lett. **65**, 998 (1994)
21. M. Yamaguchi, T. Yagi, T. Azuhata et al., J. Phys. Cond. Matter **9**, 241 (1997)
22. K. Strössner, S. Ves, M. Cardona, Phys. Rev. B **32**, 6614 (1985)
23. A.O. Awodugba, A.A. Ibiyemi, M. Tech, Pac. J. Sci. Technol. **13**, 206 (2012)
24. N.R. Murphy, J.T. Grant, L. Sun et al., Opt. Mater. **36**, 1177 (2014)
25. G.B. Sakra, I.S. Yahia, M. Fadel et al., J. Alloys Comp. **507**, 557 (2010)
26. K. Chandehari, H. Luo, A.L. Ruoff et al., Solid State Commun. **95**, 385 (1995)
27. D.T. Cromer, R.M. Douglass, E. Staritzky, Analyt. Chem. **29**, 316 (1957)
28. R.A. Penneman, E. Staritzky, J. Inorg. Nucl. Chem. **6**, 112 (1958)
29. X. Bai, M.E. Kordesch, Appl. Surf. Sci. **175–176**, 499 (2001)
30. S.S. Batsanov, Russ. J. Phys. Chem. A **90**, 250 (2016)
31. W. Faber, Z. Krist. **85**, 223 (1933)
32. K. Hattori, W. McCrone, Analyt. Chem. **28**, 1791 (1956)
33. S.F. Marenkin, A.M. Rauhkman, I.N. Matsyuk et al., Inorg. Mater. **27**, 21427 (1991). (in Russian)
34. K. Bascar, C.R. Raja, K. Thangaraj, R. Gobinathan, Mater. Chem. Phys. **28**, 1 (1991)
35. J.D. Grice, E. Gasparrini, Canad. Miner. **19**, 337 (1981)
36. J.A. Kohn, G. Katz, J.D. Broder, Amer. Miner. **42**, 398 (1957)
37. A.W. Laubengayer, H.W. McCune, J. Am. Chem. Soc. **74**, 2362 (1952)
38. I.T. Kozlov, I.P. Levshov, Proc. Miner. Soc. USSR **91**, 72 (1962). (in Russian)
39. Th Marcopoulos, M. Economou, Amer. Miner. **66**, 1020 (1981)
40. S.G. Warren, Appl. Opt. **25**, 2650 (1986)
41. M. El-Nahass, J. Mater. Sci. **27**, 6597 (1992)
42. W.Y. Liang, G. Lucovsky, R.M. White et al., Phil. Mag. **33**, 493 (1976)
43. M. Gurvitch, S. Luryi, A. Polyakov et al., J. Appl. Phys. **102**, 033504 (2007)
44. Y.Y. Guo, C.K. Kuo, P.S. Nixcholson, Solid State Ionics **110**, 327 (1998)
45. L.R. Batsanova, Proc. Sib. Div. Acad. Sci. USSR, Chem. **3**, 83 (1963) (in Russian)
46. H. Strunz, Naturwissensch. **52**, 493 (1965)
47. W.D. Birch, A. Pring, A. Reller, H.W. Schmalle, Amer. Miner. **78**, 827 (1993)
48. M.E. Thompson, C.H. Roach, R. Meyrowitz, Science **123**, 990 (1956)
49. G.A. Nazri, C. Julien, H.S. Mavi, Solid State Ionics **70/71**, 137 (1994)
50. J. Zuk, H. Kiefte, M.J. Clouter, J. Chem. Phys. **92**, 917 (1990)
51. R.M. Duglass, E. Staritzky, Anal. Chem. **29**, 315 (1957)
52. W.L. Hill, G.T. Faust, D.S. Reynolds, Amer. J. Sci. **242**, 542 (1944)
53. L. Gao, F. Lemarchand, M. Lequime, Opt. Express **20**, 15734 (2012)
54. N.E. Filonenko, I.V. Lavrov, S.V. Andreeva, Dokl. Acad. Sci. USSR **124**, 155 (1959). (in Russian)
55. T. Bååk, Appl. Opt. **21**, 1069 (1982)
56. E.E. Burovaya, Trudy Inst. Kristallogr. **5**, 197 (1949). (in Russian)
57. J.I. Larruquert, A.P. Pérez-Marín, S. García-Cortés et al., J. Opt. Soc. Am. A **29**, 117 (2012)

Chapter 5
Refractive Indices of Ternary or Complex Halides and Oxides

The data are quoted from books [1, 2], and the review [3] or measured by the authors, except where specified otherwise (Tables 5.1, 5.2, 5.3, 5.4, 5.5 and 5.6).

Table 5.1 Refractive indices of the $M_k A_l F_m$ fluorides

Compounds	n_g	n_m	n_p	Compounds	n_g	n_m	n_p
LiBeF$_3^a$		1.35		RbUF$_5^c$	1.527	?	1.512
LiBaF$_3$		1.50		RbU$_3$F$_{13}^c$	1.598	?	1.588
LiYF$_4^b$	1.472	1.454		Rb$_2$BeF$_4^d$		1.383	
LiTbF$_4$	1.5029	1.4735		Rb$_2$UF$_6^c$	1.487	1.477	1.473
LiK$_2$AlF$_6$		1.391	1.390	Rb$_2$UF$_{14}^c$	1.550	1.542	
Li$_3$InF$_6$	1.43	1.422	1.41	Rb$_2$AmF$_6^c$	1.523	1.499	1.495
Li$_3$FeF$_6$		1.42		Rb$_2$SiF$_6$		1.3534	
Li$_3$Na$_3$(AlF$_6$)$_2$		1.3395		Rb$_2$GeF$_6$		1.3961	1.3898
NaMnF$_3^e$		1.425		Rb$_3$ZrF$_7$		1.424	
NaCoF$_3$		1.455		Rb$_3$UF$_7$ ^{f}		1.438	
NaNiF$_3$		1.462		Rb$_3$MoO$_3$F$_3$		1.583	
NaCeF$_4^f$		1.514	1.493	NH$_4$BeF$_3$		1.34	
NaBF$_4$	1.3071	1.3009	1.3006	NH$_4$ZnF$_3$	1.481	1.47	
NaUF$_4^f$		1.564	1.552	NH$_3$BF$_3$	1.36	1.345	1.335
NaPuF$_4^f$		1.552	1.523	NH$_4$MnF$_3^e$		1.493	
NaK$_2$AlF$_6$		1.376		NH$_4$AlF$_4$		1.40	
Na$_2$AlF$_5$		1.349		NH$_4$FeF$_4$	1.516	1.502	1.492
Na$_2$SbF$_5$	1.476	1.467	1.435	(NH$_4$)$_2$BeF$_4^e$	1.404	1.399	1.392
NaK$_2$AlF$_6$		1.376		(NH$_4$)$_2$MnF$_5$	1.512	1.495	1.452
NaBeTh$_3$F$_{15}$		1.552	1.544	(NH$_4$)$_2$CeF$_6$ ^{g}	1.521	1.505	1.496
NaBeU$_3$F$_{15}$		1.608	1.598	(NH$_4$)$_2$SiF$_6$		1.406	1.391

(continued)

© The Author(s) 2016
S.S. Batsanov et al., *Refractive Indices of Solids*,
SpringerBriefs in Applied Sciences and Technology,
DOI 10.1007/978-981-10-0797-2_5

Table 5.1 (continued)

Compounds	n_g	n_m	n_p	Compounds	n_g	n_m	n_p
Na_2TiF_6		1.419	1.412	$(NH_4)_2GeF_6$		1.428	1.425
Na_2SiF_6		1.3124	1.3088	$(NH_4)_3AlF_6$		1.404	
Na_2GeF_6		1.3376	1.3317	$(NH_4)_3InF_6$		1.412	
Na_2SbF_5	1.476	1.467	1.435	$(NH_4)_3FeF_6$		1.442	
Na_3AlF_6	1.3387	1.3377	1.3376	$(NH_4)_3ZrF_7$		1.433	
Na_3FeF_6	1.3387	1.3377	1.3376	$(NH_4)_3HfF_7$		1.426	
$Na_5Al_3F_{14}$		1.3486	1.3424	$(NH_4)_4CeF_8^g$	1.485	1.475	1.468
$KMgF_3^h$		1.392		$(NH_4)_4UF_8^g$	1.464	1.459	1.454
$KCaF_3^h$		1.388		$(NH_4)_2MoO_3F_2$	1.886	1.702	1.668
$KCdF_3^h$		1.46		$(NH_4)_3MoO_3F_3$		1.585	
$KZnF_3$		1.53		NH_4WO_3F		1.985	
$KMnF_3$		1.447		$(NH_4)_9(WO_3)_5F_9$	1.846	1.725	1.643
$KCoF_3$		1.475		$CsMnF_3^c$	1.56	?	1.46
$KNiF_3$		1.477		$CsNiF_3$	1.5102	1.4784	
KBF_4	1.3248	1.3243	1.3113	$CsBF_4$		1.36	
K_2MgF_4		1.379	1.377	$CsThF_5^i$	1.544	?	1.528
K_2MnF_5	1.464	1.436	1.408	$CsTh_2F_9^i$		1.532	1.524
K_2TiF_6		1.475		$CsTh_3F_{13}^i$	1.558	?	1.554
K_2ZrF_6	1.498	1.465	1.454	$Cs_2ThF_6^i$	1.506	?	1.500
K_3ZrF_7		1.408		$Cs_2Th_3F_{14}^i$	1.526	?	1.518
K_2HfF_6	1.461	?	1.449	$Cs_3ThF_7^i$		1.464	
K_3HfF_7		1.403		Cs_2SiF_6		1.3847	
K_2SiF_6		1.339		Cs_2GeF_6		1.3985	
K_2GeF_6		1.383	1.381	$Cs_3MoO_3F_3$		1.57	
K_3TaF_7	1.418	?	1.414	$Cs_3(MoO_3)_2F_3$		1.680	
K_3NbF_7	1.440	?	1.437	$AgZnF_3^j$		1.620	
$KThF_5^m$		1.475	1.470	$BeCaF_4^k$	1.393	?	1.388
$KTh_2F_9^m$		1.532	1.490	$BeBaF_4^l$	1.461	1.456	1.447
$KTh_3F_{13}^m$		1.536	1.519	$BePbF_4^a$	1.620	?	1.612
$K_2ThF_6^m$		1.488	1.463	$BePb_3F_8^a$		1.725	
$K_5ThF_9^m$		1.403	1.397	$MgBaF_4$	1.4724	1.4657	1.4490
$KU_2F_9^c$	1.584	1.520		$MgZnF_4$	1.41	1.40	
$KU_6F_{25}^c$		1.595		$MgMnF_4$	1.452	1.42	
$K_2UF_6^c$	1.512	1.484		$MgCoF_4$	1.470	1.447	
$K_3UF_7^c$	1.426	?	1.410	Mg_2CaF_6	1.411	?	1.405
$K_7UF_{31}^c$		1.510		$CaZnF_4$		1.465	1.455
K_3AlF_6		1.377		$Ca_5Y_2F_{16}$		1.45	
K_3FeF_6		1.416		$SrZnF_4$		1.455	
$RbBeF_3$		1.338		$SrCoF_4$	1.496	1.492	1.477
$RbMnF_3$		1.478		$SrNiF_4$	1.500	1.497	1.479

(continued)

Table 5.1 (continued)

Compounds	n_g	n_m	n_p	Compounds	n_g	n_m	n_p
$RbCoF_3$		1.511		$BaZnF_4$	1.5214	1.5144	1.4967
$CoYF_5$	1.559	1.540	1.529	$BaMnF_4$	1.505	1.499	1.480
$CoDyF_5$	1.575	1.562	1.547	BaY_2F_8	1.5356	1.5235	1.5145
$CoErF_5$	1.573	1.560	1.546	$ZnMnF_4$	1.517	1.487	
$CoTmF_5$	1.572	1.559	1.545				

[a][4], [b][5], [c][6, 7], [d][8], [e][9, 10], [f][11], [g][12], [h][13], [i][14], [j][15], [k][16], [l][74] and [m][17]

Table 5.2 Refractive indices of ternary chlorides and oxychlorides

Compounds	n_g	n_m	n_p	Compounds	n_g	n_m	n_p
$NaMgCl_3^a$		1.604	1.586	Cs_2CuCl_4	1.678	1.648	1.625
$Na_2MgCl_4^a$	1.600	?	1.579	$Cs_2CoCl_4^b$	1.687	1.674	1.608
$Na_2CoCl_4^c$	1.58	?	1.55	$Cs_3Tl_2Cl_9$		1.784	1.774
NaK_3FeCl_6	1.5894	1.5886		Cs_4CdCl_6	1.748	1.740	
$Na_3KMnCl_6^d$	1.578	1.577		$Cu_2(OH)_3Cl^e$	1.880	1.861	1.831
$KCaCl_3$		1.52		$Cu_2(OH)_3Cl^f$	1.848	1.842	
$KPb_2Cl_5^g$	2.024	2.010	2.004	Hg_2OCl^h	2.66	2.64	2.35
K_4CdCl_6	1.5907	1.5906		Hg_2OCl_2	2.21	2.19	
K_4MnCl_6		1.59		Hg_3OCl_4		2.001	
$RbMnCl_3^i$	1.712	?	1.696	$Sr_3PuCl_9^k$		1.77	
$RbPu_2Cl_7^j$		1.80		$Ba_2MgCl_6^a$	1.726	?	1.718
Rb_4CdCl_6		1.580		$Ba_3PuCl_9^k$		1.76	
$(NH_4)_2ZnCl_4$	1.600	1.590	1.585	$PbOHCl$	2.158	2.116	2.077
$(NH_4)_3ZnCl_5$		1.538		$PbSbO_2Cl^h$	2.40	2.35	2.30
$(NH_4)_2FeCl_4$		1.644		Pb_2OCl_2		2.15	2.04
$CsCaCl_3^l$		1.603		$Pb_3O_2Cl_2$	2.31	2.27	2.24
$CsHgCl_3$		1.791		$Pb_4O_3Cl_2$	2.34	2.29	
$Cs_2HgI_4^n$		2.69		Pb_3OCl_4	2.21	2.13	
Cs_2PuCl_6		1.728	1.699	$UCl_5 \cdot PCl_5^m$	1.736	1.700	1.697
$CsPu_2Cl_7^j$		1.85					

[a][18], [b][19], [c][20], [d][21], [e]atacamite, [f]paratacamite, [g][22], [h]for λ = 671 nm, [i][23], [j][24], [k][25], [l][26], [m][27] and [n][28]

Table 5.3 Refractive indices and densities (ρ, g/cm³) of the hydrofluorides of Groups 1 and 2 metals [29, 30]

Compounds	ρ	n_g	n_m	n_p	Compounds	ρ	n_g	n_m	n_p
$LiF \cdot HF$	2.22	1.408	1.319		$CaF_2 \cdot 2HF$	2.613	1.415	1.387	1.355
$NaF \cdot HF$	2.14	1.331	1.260		$SrF_2 \cdot HF$	3.999	1.472	1.437	1.418
$KF \cdot HF$	2.36		1.352	1.327	$SrF_2 \cdot 2.5HF$	3.068	1.400	1.380	1.344
$KF \cdot 2HF$	2.06	1.315	1.311	1.305	$BaF_2 \cdot HF$	4.603	1.498	1.467	1.446
$RbF \cdot HF$	3.27		1.373	1.357	$BaF_2 \cdot 3HF$	3.560	1.425	1.411	1.334
$CsF \cdot HF$	3.86		1.414	1.410	$BaF_2 \cdot 4.5HF$	2.947		1.360	

Table 5.4 Refractive indices of the complex oxides of Group 13 metals

Compounds	n_g	n_m	n_p	Compounds	n_g	n_m	n_p
LiB_3O_5	1.6172	1.6023	1.5752	$Ca(GaO_2)_2^a$	1.778	?	1.751
$LiGeBO_4$	1.6914	1.6857		SrB_4O_7	1.7377	1.7353	1.7340
$Li_2B_4O_7^b$		1.612	1.554	$Sr(AlO_2)_2$	1.663	?	1.649
$LiAlO_2$		1.6223	1.6038	$SrAl_4O_7$	1.640	?	1.614
$LiAl_5O_8$		1.725		$SrAl_{12}O_{19}$		1.702	1.694
$LiGaO_2$	1.7654	1.7350		$Sr_3(AlO_3)_2$		1.728	
$NaBO_2^c$		1.570	1.461	$SrAl_{12}O_{19}$		1.702	1.694
$NaB_3O_5^d$	1.583	1.530	1.518	$SrLaAlO_4$	1.9865	1.9663	
$Na_2B_4O_7^d$	1.570	1.535	1.515	$Ba(BO_2)_2$		1.676	1.528
$NaAlO_2$	1.580	1.575	1.566	$Ba_3(BO_3)_2$	1.768	?	1.756
$Na_2Al_{12}O_{19}$		1.686	1.650	$Ba(AlO_2)_2$		1.683	
KBO_2^c		1.526	1.450	$BaAl_{12}O_{19}$		1.702	1.694
$KAlO_2$		1.603		$Ba_3(AlO_3)_2$		1.735	
$Na_2Al_{12}O_{19}$		1.696	1.660	$Zn(BO_2)_2^e$	1.673	?	1.643
$KNb(BO_3)_2$	1.806	1.777		$Zn_3(BO_3)_2^e$	1.720	?	1.669
$RbNb(BO_3)_2$	1.802	1.777	1.757	$Zn_4O(BO_2)_6$		1.7464	
CsB_3O_5	1.5892	1.5615	1.5308	$Zn(AlO_2)_2$		1.805	
$Cu(BO_2)_2^f$		1.75		$Zn(GaO_2)_2$		1.74	
$Cu_2BO(OH)^g$	1,769	1.699	1.627	$Sc_2B_2O_5F_2^h$		1.598	
Be_2BO_3OH	1.631	1.591	1.560	$LaBO_3^H$	1.882	1.856	1.800
$Be(AlO_2)_2$	1.753	1.747	1.744	$La_3BO_6^H$	1.953	1.948	1.907
$BeMg(AlO_2)_4$		1.724	1.716	$LaB_3O_6^H$	1.791	1.769	1.694
$MgHBO_3$	1.60	1.575		$NdBO_3$	1.903	1.818	
$MgAlBO_4^i$	1.708	1.700	1.672	$Nd(BO_2)_3$	1.800	1.780	1.725
$Mg(BO_2)_2^j$	1.660	?	1.605	$GdBO_3$	1.831	1.822	
$Mg(B_3O_5)_2^k$	1.504	1.500	1.442	$ErBO_3$	1.825	1.816	
$Mg_2B_2O_5^l$	1.674	1.660	1.589	$YbBO_3$	1.822	1.812	
$Mg_3(BO_3)_2$	1.675	1.654	1.653	$AlBO_3$		1.653	1.640
$Mg_3BO_3F_3$		1.5086	1.4858	$InBO_3$		1.878	1.776
$Mg_3B_7O_{13}Cl$		1.6713		Al_3BO_6	1.623	1.603	1.586
$Mg(AlO_2)_2$		1.722		$Y_3Al_5O_{12}$		1.8352	
$CaHBO_3^i$	1.658	1.643	1.555	$Dy_3Al_5O_{12}$		1.8613	
$Ca(BO_2)_2$	1.682	1.656	1.540	$Ho_3Al_5O_{12}$		1.8567	
CaB_4O_7	1.638	1.568		$Er_3Al_5O_{12}$		1.8522	
$CaB_2O_2(OH)_4^m$		1.614	1.585	$Lu_3Al_5O_{12}$		1.8423	
$Ca_2B_2O_5$	1.667	1.662	1.585	$Y_3Ga_5O_{12}$		1.919	
$Ca_2B_4O_7^o$	1.577	?	1.574	$Gd_3Ga_5O_{12}$		1.9698	
$Ca_3(BO_3)_2^p$		1.726	1.630	$Tb_3Ga_5O_{12}$		1.971	
$Ca_3B(OH)_6^q$	1.570	1.568	1.557	$GaAlO_3^n$		1.69	
$CaB_2O_2(OH)_4^m$	1.614	~1.614	1.585	PbB_4O_7	1.936	1.934	1.930
$Ca_2CuB_2(OH)_{12}^s$	1.615	1.608	1.585	$Pb(AlO_2)_2$		1.91	1.85
$Ca_3B(OH)_6^q$	1.570	1.568	1.557	$ThB_2O_5^r$	1.823	1.750	1.720

(continued)

Table 5.4 (continued)

Compounds	n_g	n_m	n_p	Compounds	n_g	n_m	n_p
Ca₃(BO₃)₂[t]		1.726	1.630	Bi(BO₂)₃	1.9518	1.8111	1.7806
CaB₃O₅OH[u]	1.650	1.637	1.608	MnSn(BO₃)₂[t]		1.854	1.757
Ca(AlO₂)₂	1.682	1.656	1.540	Mn₃B₄O₉	1.776	1.738	1.617
CaAl₄O₇	1.651	1.617	1.617	Mn₃B₇O₁₃Cl[v]	1.744	1.737	1.732
CaAl₁₂O₁₉		1.757	1.750	Mn(AlO₂)₂		1.848	
Ca₃(AlO₃)₂		1.710		Fe(AlO₂)₂		1.83	
CaYAlO₄	1.9290	1.9077		Co₂B₂O₅[w]	1.86	?	1.82
CaNdAlO₄	1.9867	1.9620		Co₃(BO₃)₂[w]	1.89	?	1.84
				Ni(AlO₂)₂[x]		1.875	

[a][31], [b][32], [c][33], [d][34], [e][35], [f][36], [g][37], [h][38], [H][39], [i][40], [j][41], [k][42], [l][43], [m][44], [n][45], [o][46], [p][47], [q][48], [r][49], [s][50], [t][51], [u][52], [v][53], [w][54] and [x][55]

Table 5.5 Refractive indices of complex oxides of Ti, V, Nb and Ta

Compounds	n_g	n_m	n_p	Compounds	n_g	n_m	n_p
Li₂TiO₃		2.087		SrVO₃	1.99	?	1.93
Li₂Ti₃O₇	2.4297	2.3513	2.1600	Sr₅(VO₄)₃F		1.8576	1.8416
MgTiO₃		2.31	1.95	BaV₂O₅	2.05	2.00	1.98
MgTi₂O₅	2.23	2.19	2.11	YVO₄	2.2285	2.0022	
Mg₂TiO₄		1.959		CeV₂O₇*[a]	2.29	2.01	
CaTiO₃		2.407	2.210	CeVO₃*[a]	2.23	2.01	
Ca₃Ti₂O₇[b]	2.22	?	2.16	BiVO₄*	2.51	2.50	2.41
SrTiO₃[c]		2.4089		Bi₃O(VO₄)₂OH[d]		2.42	
BaTiO₃		2.4408	2.3833	Mn₅(VO₄)₂(OH)₄[e]	1.810	?	1.803
Ba₂TiO₄	1.954	?	1.891	Mn₇(VO₄)₂(OH)₈[f]	1.77	1.762	1.74
Zn₂TiO₄		2.17		LiNbO₃[g]		2.286	2.202
PbTiO₃[c]		2.6977		Li₂K₃(NbO₃)₅		2.2954	2.1755
MnTiO₃		2.481	2.210	NaNbO₃[h]		2.30	
Al₂TiO₅[i]	2.06	?	2.025	NaBa₂(NbO₃)₅	2.3389	2.3376	2.2295
TiTi₂O₅	2.32	?	2.19	KNbO₃[j]		2.1815	
Bi₃TiO₅		2.5873		KPb₂(NbO₃)₅	2.4623	2.4486	2.3883
Bi₄Ti₃O₁₂[c]		2.6725		Sr₂Nb₂O₇	2.17	2.16	2.06
Fe₂TiO₅*	2.42	2.39	2.38	Ba₃La(NbO₄)₃		2.270	2.104
KVO₃[k]	1.901	1.842	1.720	Pb(NbO₃)₂*[l]	2.60	2.43	2.40
K₄Mn(VO₄)₃[m]	1.988	1.960	1.925	LaNbO₄	2.200	2.105	2.087
NH₄VO₃[k]		1.927	1.768	VNb₂O₇*[a]	2.28	?	2.22
β-Cu₂V₂O₇[n]		2.055		LiTaO₃	2.1902	2.1859	
Cu₃Pb(VO₄)₂Cl₂[o]	2.35	2.29		KTaO₃		2.2425	
Ag₄V₂O₇[k]		2.17		Sn(TaO₃)₂	2.499	2.418	2.388
Mg₃V₆O₁₃[a]	2.56	2.10	1.99	Mn(TaO₃)₂	2.34	2.25	2.19
Ca₃(VO₄)₂		1.8919	1.8682	Ba₃La(TaO₄)₃		2.161	2.021

[a][56], [b][57], [c][58], [d][59], [e][60], [f][61], [g][62], [h][63], [i][64], [j][65], [k][66], [l][67], [m][68], [n][69] and * indicates at λ = 671 nm

Table 5.6 Refractive indices of the complex oxides of Cr, Mo, W, Mn, Fe (tet – tetragonal and mon – monoclinic phase)

Compounds	n_g	n_m	n_p	Compounds	n_g	n_m	n_p
HCrO$_2^a$		2.155	1.975	Tl$_2$WO$_4$		2.08	
Na$_2$Cr$_2$O$_7$	1.747	1.708	1.659	CaWO$_4$	1.9365	1.9200	
NaK$_3$(CrO$_4$)$_2$	1.736	1.728		tet-PbWO$_4$		2.27	2.19
K$_2$CrO$_4$	1.730	1.726	1.709	mon-PbWO$_4$	2.30	2.27	2.27
K$_2$Cr$_2$O$_7$	1.818	1.742	1.725	MnWO$_4$	2.283	2.195	2.150
(NH$_4$)$_2$Cr$_2$O$_7$	1.864	1.783	1.716	FeWO$_4$	2.414	2.305	2.255
Rb$_2$Cr$_2$O$_7$	1.824	1.740	1.712	KY(WO$_4$)$_2$		1.889	
MgCr$_2$O$_4$		2.08		KPr(WO$_4$)$_2$		1.896	
FeCr$_2$O$_4$		2.12		KNd(WO$_4$)$_2$		1.857	
NiCr$_2$O$_4$		2.29		KMnO$_4$*	1.85	?	1.80
Cu$_3$(MoO$_4$)$_2$(OH)$_2$	2.020	2.002	1.930	ZnMn$_2$O$_4$		2.35	2.10
CaMoO$_4$	1.984	1.974		MnMn$_2$O$_4$*		2.45	2.15
SrMoO$_4$	1.926	1.921		LiFeO$_2^b$		2.14	
PbMoO$_4$	2.4053	2.2826		MgFe$_2$O$_4$		2.39	
Y$_2$(MoO$_4$)$_3$		2.031		CaFe$_2$O$_4$	2.58	2.43	
Ce$_2$(MoO$_4$)$_3$		2.0403	2.0277	Ca$_2$Fe$_2$O$_5$*	2.29	2.22	2.21
Pr$_2$(MoO$_4$)$_3$		2.007		ZnFe$_2$O$_4$*		2.36	
Nd$_2$(MoO$_4$)$_3$		2.0239	2.0218	CdFe$_2$O$_4$*		2.39	
KPr(MoO$_4$)$_2$		1.900		MnFe$_2$O$_4$*		2.3	
KNd(MoO$_4$)$_2$		1.898		FeFe$_2$O$_4$		2.42	
KEr(MoO$_4$)$_2$		1.877		CoFe$_2$O$_4$		2.48	
H$_2$WO$_4^c$	2.04	?	1.82	NiFe$_2$O$_4$		2.69	
NaAl(WO$_4$)$_2$	1.702	?	1.658	Y$_3$Fe$_5$O$_{12}$		2.3266	
(NH$_4$)$_2$WO$_4^d$		1.709					

[a][70], [b][71], [c][72], [d][73] and * indicates at $\lambda = 671$ nm

References

1. E. Kordes, *Optische Daten* (Verlag Chemie, Weinheim, 1960)
2. A.N. Winchell, H. Winchell, *Optical properties of artificial minerals* (Academic Press, New York and London, 1964)
3. R.D. Shannon, R.C. Shannon, O. Medenbach, R.X. Fischer, J. Phys. Chem. Ref. Data **31**, 931 (2002)
4. D.M. Roy, R. Roy, E.F. Osborn, J. Am. Ceram. Soc. **37**, 300 (1954)
5. R.E. Thoma, C.F. Weaver, H.A. Friedman et al., J. Phys. Chem. **65**, 1096 (1961)
6. R.E. Thoma, H. Insley, B.S. Landau et al., J. Am. Ceram. Soc. **41**, 538 (1958)
7. F.H. Kruse, L.B. Asprey, Inorg. Chem. **1**, 137 (1962)
8. N.A. Toropov, R.G. Grebenshchikov, Rus. J. Inorg. Chem. (in Russian) **1**, 1619 (1956)
9. L.R. Batsanova, A.V. Novoselova, Yu.P. Simanov, Rus. J. Inorg. Chem. (in Russian) **1**, 2638 (1956)

10. Yu.P. Simanov, L.R. Batsanova, L.M. Kovba, ibid, **2**, 2410 (1957)
11. C.J. Barton, J.D. Redman, R.A. Strehlow, J. Inorg. Nucl. Chem. **20**, 45 (1961)
12. R.A. Penneman, A. Rosenzweig, Inorg. Chem. **8**, 627 (1969)
13. C. Brisi, Ann. Chim. (Roma) **42**, 356 (1952)
14. R.E. Thoma, T.S. Carlton, J. Inorg. Nucl. Chem. **17**, 88 (1961)
15. R.C. De Vries, R. Roy, J. Amer. Chem. Soc. **75**, 2479 (1953)
16. L.M. Mikheeva, A.V. Novoselova, R. Bikhtimirov, Rus. J. Inorg. Chem. (in Russian) **1**, 499 (1956)
17. W.J. Asker, E.R. Segnit, A.W. Wylie, J. Chem. Soc. 4470 (1952)
18. N.V. Bondarenko, Rus. J. Inorg. Chem. (in Russian) **7**, 1389 (1962)
19. M.A. Poray-Koshits, Crystallogr. (in Russian) **1**, 291 (1956)
20. K.A. Bol'shakov, P.I. Fedorov, G.D. Agashkina, Rus. J. Inorg. Chem. (in Russian) **2**, 1115 (1957)
21. A.R. Kampf, S.J. Mills, F. Nesto et al., Am. Miner. **98**, 231 (2013)
22. J. Schluter, D. Pohl, S. Britvin, Ns. Jb. Miner. Abh. **182**, 95 (2005)
23. S. Jeong, S. Haussühl, Cryst. Res. Technol. **26**, 739 (1991)
24. R. Benz, R.M. Douglass, J. Phys. Chem. **65**, 1461 (1961)
25. K.W.R. Johnson, M. Kahn, J.A. Leary, J. Phys. Chem. **65**, 2226 (1961)
26. V.E. Plyuschev, V.B. Tulinova, G.P. Kuznetsova et al., Rus. J. Inorg. Chem. (in Russian) **2**, 2212 (1957)
27. R.E. Panzer, J.F. Suttle, J. Inorg. Nucl. Chem. **20**, 229 (1961)
28. A.A. Lavrent'ev, B.V. Gabrelian, V.T. Vu, et al., Opt. Mater. **42**, 351 (2015)
29. E.D. Ruchkin, A.A. Opalovsky, E.D. Fedotova, Proc. Sib. Div. Acad. Sci. USSR Chem. (in Russian) **9**, No 4, 22 (1968)
30. E.D. Ruchkin, D.D. Ikrami, N.S. Nikolaev, Dokl. Acad. Sci. USSR (in Russian) **174**, 1349 (1967)
31. J. Jeevaratnam, F.P. Glasser, J. Am. Ceram. Soc. **44**, 563 (1961)
32. D. London, M.E. Zolensky, E. Roedder, Canad. Miner. **25**, 173 (1987)
33. N.A. Toropov, Dokl. Acad. Sci. USSR (in Russian) **23**, 74 (1939)
34. S.S. Cole, S.R. Scholes, C.R. Amberg, J. Am. Ceram. Soc. **18**, 58 (1935)
35. J. Schluer, D. Pohl, U. Golla-Achindler, Chile. N. Jb. Miner. Abh. **185**, 27 (2008)
36. N.A. Toropov, P.F. Konovalov, Dokl. Acad. Sci. USSR (in Russian) **66**, 1105 (1949)
37. N.N. Pertsev, W. Schreyer, Th Armbruster et al., Eur. J. Miner. **16**, 151 (2004)
38. L.R. Batsanova, L.A. Novosel'tseva, A.I. Madaras, Inorg. Mater. (in Russian) **10**, 621 (1974)
39. E.M. Levin, C.R. Robbins, J.L. Waring, J. Amer. Ceram. Soc. **44**, 87 (1961)
40. G.F. Claringbull, M.H. Hey, Miner. Mag. **29**, 841 (1952)
41. V.G. Hill, R. Roy, E.F. Osborn, J. Amer. Ceram. Soc. **35**, 135 (1952)
42. K. Walenta, Tscherm. Min. Petrog. **26**, 69 (1979)
43. Y. Baskin, Y. Harada, J.H. Handwerk, J. Amer. Ceram. Soc. **44**, 456 (1961)
44. D.P. Shashkin, M.A. Simonov, N.I. Chernova, et al., Dokl. Acad. Sci. USSR (in Russian) **182**, 1402 (1968)
45. N.N. Vasil'kova, Proc. Miner. Soc. USSR (in Russian) **91**, 455 (1962)
46. S.V. Malinko, B.P. Fitsev, N.N. Kuznetsova et al., Proc. Miner. Soc. USSR (in Russian) **109**, 469 (1980)
47. S.I. Kovalenko, A.V. Voloshin, Ya.A. Pakhomovsky, et al., Dokl. Acad. Sci. USSR (in Russian) **272**, 1449 (1983)
48. M.A. Bogomolov, I.B. Nikitina, N.N. Pertsev, Dokl. Acad. Sci. USSR (in Russian) **184**, 1398 (1969)
49. P.F. Konovalov, Dokl. Acad. Sci. USSR (in Russian) **70**, 847 (1950)
50. I. Nakai, H. Okada, K. Masutomi et al., Am. Miner. **71**, 1234 (1986)
51. I. Kusachi, C. Henmi, S. Kobayashi, Miner. Mag. **59**, 549 (1995)
52. H. Gaertner, K.L. Roese, R. Kühn, Naturwiss. **49**, 230 (1962)
53. R.M. Honea, F.R. Beck, Am. Miner. **47**, 665 (1962)
54. H.M. Davis, M.A.Knight, J. Am. Ceram. Soc. **28**, 97 (1945)

55. N.L. Dilaktorsky, Proc. Miner. Soc. USSR (in Russian) **68**, 18 (1939)
56. B.W. King, L.L. Suber, J. Am. Ceram. Soc. **38**, 306 (1955)
57. H.G. Fisk, J. Am. Ceram. Soc. **34**, 9 (1951)
58. M. Simon, F. Mersch, C. Kuper, et al., Phys. Stat. Solidi a**159**, 559 (1997)
59. K. Walenta, P.J. Dunn, G. Hentschel G., et al., Tscher. Min. Petrogr. Mitt. **31**, 165 (1983)
60. R. Basso, G. Lucchetti, L. Zefiro, A. Palenzona. Z. Krist. **201**, 223 (1992)
61. J. Brugger, P. Elliott, N. Meisser, S. Ansermet, Am. Miner. **96**, 1894 (2011)
62. H. Han, L. Cai, H. Hu, Opt. Mater. **42**, 47 (2015)
63. A. Safiankoff, Bull. Acad. Roy. Soc. **5**, 1251 (1959)
64. V.A. Bron, A.K. Podnogin, Dokl. Acad. Sci. USSR (in Russian) **91**, 93 (1953)
65. N. Umemura, K. Yoshida, K. Kato, Appl. Opt. **38**, 991 (1999)
66. S.S. Batsanov, A.S. Sonin, Crystallography (in Russian) **1**, 321 (1956)
67. M.H. Francombe, B. Lewis, Acta Cryst. **11**, 696 (1958)
68. T. Witzke, F. Zhen, K. Seff et al., Am. Miner. **86**, 1081 (2001)
69. J.M. Hughes, R.W. Birnie, Am. Miner. **65**, 1146 (1980)
70. B.J. Thamer, R.M. Douglass, E. Staritzky, J. Am. Chem. Soc. **79**, 547 (1957)
71. D.W. Strickler, R. Roy, J. Amer. Ceram. Soc. **44**, 225 (1961)
72. P.F. Kerr, F.Young, Am. Miner. **29**, 192 (1944)
73. L.N. Formozova, Rus. J. Gen. Chem. (in Russian) **15**, 863 (1945)
74. D.F. Kirkina, A.V. Novoselova, Yu.P. Simanov, Rus. J. Inorg. Chem. (in Russian) **1**, 125 (1956)

Chapter 6
Refractive Indices of Silicates and Germanates

This section describes the optical properties of silicates and germanates, which comprise no more than two types of cations and/or two additional anions. Only compounds of rational composition (daltonides) are included, as the RIs of solid solutions of variable compositions can be calculated by interpolation (Tables 6.1, 6.2, 6.3, 6.4, 6.5).

Table 6.1 Refractive indices of orthosilicates (anion SiO_4^{2-})

Compounds	n_g	n_m	n_p	Compounds	n_g	n_m	n_p
Li_4SiO_4	1.614	1.60	1.594	$Ca_3Ga_2(SiO_4)_3$		1.771	
$LiLaSiO_4$		1.870	1.843	$Ca_3In_2(SiO_4)_3$		1.775	
$LiAlSiO_4^a$	1.586	1.578	1.575	$Ca_3V_2(SiO_4)_3$		1.882	
$LiGaSiO_4^b$		1.548		$Ca_3Cr_2(SiO_4)_3$		1.852	
$Li_2BeSiO_4^c$	1.633	1.628	1.622	$Ca_3Fe_2(SiO_4)_3$		1.897	
$NaYSiO_4$		1.832	1.804	Ca_3OSiO_4		1.722	1.716
$NaLaSiO_4$		1.867	1.840	$Ca_5(SiO_4)_2F_2^d$	1.608	1.605	1.594
$NaPrSiO_4^g$		1.8891	1.861	$Ca_8Mg(SiO_4)_4Cl_2^e$		1.676	
$NaNdSiO_4^g$		1.889	1.861	$Ca_9(SiO_4)_4F_2^f$	1.631	1.625	1.621
$NaSmSiO_4^g$		1.898	1.8967	$Ca_9(SiO_4)_4(OH)_2^f$	1.635	1.629	1.625
$NaAlSiO_4^i$	1.537	1.534		Sr_2SiO_4	1.756	1.732	1.727
$NaGaSiO_4^b$	1.629	1.623		$SrB_2(SiO_4)_2^h$	1.632	1.627	1.597
Na_2CaSiO_4		1.60		Ba_2SiO_4	1.830	?	1.810
Na_2MgSiO_4		1.523		$BaB_2(SiO_4)_2^h$	1.656	1.656	1.649
$Na_2TiOSiO_4^k$	1.765	1.741	1.740	$BaAl_2(SiO_4)_2$	1.600	1.593	1.587
Na_4SiO_4	1.537	?	1.524	$Ba_2Ti(SiO_4)_2^j$		1.775	1.765
$Na_4Zr_2(SiO_4)_3$		1.715	1.692	Zn_2SiO_4	1.723	1.689	
Na_6OSiO_4	1.529	?	1.524	$ZnPbSiO_4$	1.96	1.95	1.92
$Na_2ZrOSiO_4$	1.790	?	1.741	$Cd_3Al_2(SiO_4)_3$		1.817	

(continued)

© The Author(s) 2016
S.S. Batsanov et al., *Refractive Indices of Solids*,
SpringerBriefs in Applied Sciences and Technology,
DOI 10.1007/978-981-10-0797-2_6

Table 6.1 (continued)

Compounds	n_g	n_m	n_p	Compounds	n_g	n_m	n_p
$KLaSiO_4$		1.867	1.840	$Cd_3In_2(SiO_4)_3$		1.764	
$KAlSiO_4^n$		1.540	1.535	Hg_2SiO_4	1.95	1.89	
$KGaSiO_4^b$	1.579	1.569		$Sc_2OSiO_4^l$	1.850	?	1.835
K_2CaSiO_4	1.605	1.600		$Y_2OSiO_4^m$	1.825	?	1.807
K_2ZnSiO_4		1.622		$Y_4(SiO_4)_3^m$		1.780	1.765
$RbAlSiO_4^g$		1.530	1.526	$Y_5(SiO_4)_3(OH)_3^o$	1.827	1.827	1.786
$RbGaSiO_4^b$		1.576		$La_2OSiO_4^p$	1.875	1.855	
$CsAlSiO_4^g$		1.574		$La_4(SiO_4)_3^{3p}$		1.852	1.837
$CsGaSiO_4^b$	1.534	1.528		$Ce_4(SiO_4)_3^p$		1.850	
$TlAlSiO_4^r$		1.78	1.755	$Sm_2O(SiO_4)_3^p$	1.885	?	1.863
Be_2SiO_4	1.670	1.654		$Sm_4(SiO_4)_3^p$		1.860	1.840
Mg_2SiO_4	1.669	1.651	1.636	$Gd_4(SiO_4)_3^q$		1.870	1.855
$MgCaSiO_4$	1.682	1.677	1.666	$Dy_2OSiO_4^q$	1.865	?	1.847
$MgCa_3(SiO_4)_2$	1.724	1.712	1.706	$Dy_4(SiO_4)_3^q$		1.853	1.838
$Mg_3Al_2(SiO_4)_3$		1.749		$Er_2OSiO_4^q$	1.825	?	1.807
$Mg_3F_2SiO_4$	1.590	1.567	1.561	$Er_4(SiO_4)_3^q$		1.800	1.780
α-Ca_2SiO_4	1.738	1.724		$Yb_4(SiO_4)_3^p$		1.800	1.782
β-Ca_2SiO_4	1.730	1.715	1.707	α-Al_2OSiO_4	1.680	1.664	1.658
γ-Ca_2SiO_4	1.654	1.645	1.642	β-Al_2OSiO_4	1.644	1.639	1.633
$CaBSiO_4OH$	1.669	1.653	1.625	$Al_2F_2SiO_4$	1.638	1.631	1.629
$CaAlSiO_4OH^t$	1.730	1.725	1.700	Pb_2SiO_4	2.18	2.15	2.13
$CaMnSiO_4$	1.736	1.723	1.686	$Pb_4O_2SiO_4$	2.38	2.34	2.31
$CaFeSiO_4$	1.743	1.734	1.696	$ZrSiO_4$	1.984	1.926	
$CaTiOSiO_4$	2.093	?	1.901	$ThSiO_4^s$	1.922	1.900	1.898
$CaSnOSiO_4^v$	1.799	1.784	1.765	$Bi_4(SiO_4)_3$		2.029	
$CaLa_2(SiO_4)_2^g$		1.880	1.874	Mn_2SiO_4	1.816	1.806	1.774
$CaNd_2(SiO_4)_2^g$		1.903	1.898	$Mn_3Al_2(SiO_4)_3$		1.800	
$CaB_2(SiO_4)_2$	1.636	1.633	1.630	$Mn_4As(SiO_4)_3OH^u$	1.760	1.751	1.745
$CaAl_2(SiO_4)_2^w$		1.580	1.575	$Mn_9(SiO_4)_4(OH)_2$	1.789	1.783	1.772
$Ca_3Sc_2(SiO_4)_3$		1.778		Fe_2SiO_4	1.878	1.866	1.825
$Ca_3Al_2(SiO_4)_3$		1.735		$Fe_3Al_2(SiO_4)_3$		1.830	
				Ni_2SiO_4	2.019	1.987	1.976

[a][1], [b][2], [c][3], [d][4], [e][5], [f][6], [g][7], [h][8], [i][9], [j][10], [k][11], [l][12], [m][13], [n][14], [o][15], [p][16], [q][17], [r][18], [s][19], [t][20], [u][21], [v][22], [w][23]

Table 6.2 Refractive indices of metasilicates (anion SiO_3^{2-})

Compounds	n_g	n_m	n_p	Compounds	n_g	n_m	n_p
$HNaCa_2(SiO_3)_3$	1.632	1.604	1.595	$Be_3Al_2(SiO_3)_6$		1.602	1.594
$HCa_2Fe_2(SiO_3)_5$	1.753	1.731	1.720	$MgSiO_3$	1.658	1.653	1.650
Li_2SiO_3	1.610	1.591		$MgCa(SiO_3)_2$	1.7024	1.6795	1.6726
$LiNaSiO_3$	1.571	1.557	1.552	$Mg_2Al_4O_3(SiO_3)_5$	1.541	?	1.537
$\beta\text{-}LiAl(SiO_3)_2^a$	1.524	1.519		$Mg_2Ca_5O(SiO_3)_6$	1.635	1.627	1.621
$\gamma\text{-}LiAl(SiO_3)_2^a$	1.522	1.516		$CaSiO_3$	1.654	1.611	1.610
$Li_4K_{10}(SiO_3)_7$		1.540		$Ca_2Ba(SiO_3)_3$		1.681	1.668
Na_2SiO_3	1.528	1.520	1.513	$CaZn(SiO_3)_2^b$	1.70	1.69	1.68
$Na_2Ca_2(SiO_3)_3$	1.599	?	1.596	$CaMn(SiO_3)_2$	1.738	1.719	1.710
$Na_4Ca(SiO_3)_3$		1.571		$CaFe(SiO_3)_2$	1.7551	1.7318	1.7260
$Na_8Ca_3O_2(SiO_3)_5$		1.620		$CaMn_4(SiO_3)_5$	1.739	1.731	1.729
$NaAl(SiO_3)_2$	1.667	1.659	1.655	$Ca_2Zr(SiO_3)_4^c$	1.658	?	1.653
$NaCr(SiO_3)_2^d$	1.764	1.756	1.744	$SrSiO_3$	1.637	1.599	
$Na_2Ti_2O_3(SiO_3)_2$	2.02	2.01	1.91	$BaSiO_3$	1.678	1.674	1.673
$NaFe(SiO_3)_2$	1.8271	1.8103	1.7710	$BaTi(SiO_3)_3$	1.804	1.757	
$Na_{10}Fe_2(SiO_3)_8$	1.625	1.609		$BaZr(SiO_3)_3^e$	1.691	1.681	
K_2SiO_3	1.528	1.521	1.520	$ZnSiO_3$	1.623	?	1.616
$KAl(SiO_3)_2^f$		1.508		$Al_6O_7(SiO_3)_2$	1.654	1.644	1.642
$K_2Be_3(SiO_3)_4$		1.523		$PbSiO_3$	1.968	1.961	1.947
$K_4Ca(SiO_3)_3$		1.572		$MnSiO_3$	1.739	1.735	1.733
$RbAl(SiO_3)_2^f$		1.526		$Mn_2Al_4O_3(SiO_3)_5$	1.558	1.558	1.537
$CsAl(SiO_3)_2^f$		1.514		$Fe_2Al_4O_3(SiO_3)_5$	1.574	1.564	1.551
$NH_4Al(SiO_3)_2^g$		1.518					

[a][24], [b][25], [c][26], [d][27], [e][28], [f][29], [g][30]

Table 6.3 Refractive indices of layered silicates (anion $Si_2O_5^{2-}$)

Compounds	n_g	n_m	n_p	Compounds	n_g	n_m	n_p
$HKSi_2O_5$	1.535	1.501	1.495	$K_8Ca(Si_2O_5)_5$		1.548	1.537
$LiKSi_2O_5$	1.540		1.536	$K_2Pb(Si_2O_5)_2$	1.650	1.612	1.590
$Li_2Si_2O_5$	1.558	1.550	1.547	$Rb_2Si_2O_5^a$	1.527	1.514	1.507
$LiAl(Si_2O_5)_2$	1.516	1.510	1.504	$Cs_2Si_2O_5^b$	1.544	1.538	1.532
$NaKMn(Si_2O_5)_2^d$	1.557	1.551	1.540	$MgBa(Si_2O_5)_2^c$	1.585	?	1.573
$NaCa_2(Si_2O_5)_2F^e$	1.581	1.579	1.567	$Mg_3(OH)_4Si_2O_5$	1.555	1.543	1.542
$Na_2Si_2O_5$	1.515	1.510	1.500	$Mg_3(OH)_2(Si_2O_5)_2$	1.575	1.575	1.540
$Na_2Mg_2(Si_2O_5)_3$	1.546	1.542	1.540	$CaCu(Si_2O_5)_2^f$		1.635	1.605
$K_2Si_2O_5$	1.513	1.509	1.503	$SrCu(Si_2O_5)_2^f$		1.628	1.588
$K_2Mg_5(Si_2O_5)_6$	1.550	1.543		$BaCu(Si_2O_5)_2^f$		1.632	1.593
$K_2Ca_2(Si_2O_5)_3$	1.590		1.575	$BaSi_2O_5$	1.624	1.615	1.597
$K_4Ca(Si_2O_5)_3$	1.543	1.541	1.535	$BaFe(Si_2O_5)_2$		1.621	1.619
				$Fe_3(OH)_2(Si_2O_5)_2$	1.618	1.618	1.586

[a][31], [b][32], [c][33], [d][34], [e][35], [f][36]

Table 6.4 Refractive indices of group silicates (anion $Si_2O_7^{6-}$)

Compounds	n_g	n_m	n_p	Compounds	n_g	n_m	n_p
$LiK_5Si_2O_7$	1.520	?	1.515	$Ca_2FeSi_2O_7$		1.690	1.673
$Na_2Mg_2Si_2O_7$	1.654	1.646	1.641	$Ca_3Si_2O_7$	1.650	1.645	1.641
$Na_2Ca_2Si_2O_7$		1.665		$Ca_4Si_2O_7F_2$	1.606	1.595	1.592
$Na_2CaZrSi_2O_7F_2^a$	1.639	1.634	1.627	$Ca_4Si_2O_7(OH)_2$	1.598	1.589	1.586
$Na_2ZrSi_2O_7$	1.710	?	1.688	$Sr_2CoSi_2O_7$	1.6766	1.6487	
$Na_6Si_2O_7$	1.529	?	1.524	$Ba_2FeSi_2O_7^b$	1.760	1.740	1.740
$K_2ZrSi_2O_7^c$	1.715	1.715	1.665	$Hg_6Si_2O_7^d$	2.58	?	2.10
$K_2Pb_2Si_2O_7$		1.93	1.72	$Sc_2Si_2O_7^e$	1.803	1.785	1.754
$CaZrSi_2O_7^f$	1.738	1.736	1.720	$Y_2Si_2O_7^g$	1.744	1.738	1.731
$Ca_2BeSi_2O_7^h$		1.672	1.664	$La_2Si_2O_7^i$	1.762	?	1.752
$Ca_2MgSi_2O_7$	1.6431	1.6391		$Sm_2Si_2O_7^i$	1.775	?	1.765
$Ca_2ZnSi_2O_7$		1.6735	1.6618	$Yb_2Si_2O_7^i$	1.770	?	1.740
$Ca_2CoSi_2O_7$	1.6764	1.6197		$Ce_2Si_2O_7^j$		1.770	

a[37], b[38], c[39], d[40], e[12], f[41], g[13], h[42], i[16], j[43]

Table 6.5 Refractive indices of germanates

Compounds	n_g	n_m	n_p	Compounds	n_g	n_m	n_p
$NaAlGe_3O_8$	1.619	?	1.606	$Ca_3Fe_2(GeO_4)_3$		1.962	
$NaGaGe_3O_8$	1.654	?	1.638	$Sr_3In_2(GeO_4)_3$		1.859	
$Na_2Ge_4O_9^a$	1.695	?	1.690	$Sr_3Ga_2Ge_4O_{14}$	1.8166	1.7984	
$Na_4Ge_9O_{20}$	1.699	1.688		$Ba_2ZnGe_2O_7$	1.764	1.752	
$KAlGe_3O_8$	1.595	?	1.590	Zn_2GeO_4	1.802	1.769	
$KGaGe_3O_8$	1.628	?	1.615	$Cd_3Sc_2(GeO_4)_3$		1.974	
$K_2Ge_4O_9$	1.7591	1.7273		$Cd_3Al_2(GeO_4)_3$		1.908	
$MgGeO_3^b$	1.759	1.755	1.741	$Cd_3Ga_2(GeO_4)_3$		1.960	
oliv-$Mg_2GeO_4^b$	1.765	1.717	1.698	$Cd_3In_2(GeO_4)_3$		1.990	
spin-$Mg_2GeO_4^c$		1.768		$Cd_3Ti_2(GeO_4)_3$		2.09	
$CaAl_2(GeO_4)_2$	1.647	?	1.641	$Cd_3V_2(GeO_4)_3$		2.048	
$CaGa_2(GeO_4)_2$	1.711	?	1.705	$Cd_3Cr_2(GeO_4)_3$		2.061	
Ca_2GeO_4	1.734	?	1.724	$Cd_3Mn_2(GeO_4)_3$		2.11	
$Ca_3Sc_2(GeO_4)_3$		1.84		$Cd_3Fe_2(GeO_4)_3$		2.09	
$Ca_3Al_2(GeO_4)_3$		1.796		$Pb_5Ge_3O_{11}^d$	2.1662	2.1304	
$Ca_3Ga_2(GeO_4)_3$		1.846		$Mn_3Al_2(GeO_4)_3$		1.891	
$Ca_3Ga_2Ge_4O_{14}$	1.8239	1.7996		$Mn_3Ga_2(GeO_4)_3$		1.930	
$Ca_3In_2(GeO_4)_3$		1.844		$Mn_3Cr_2(GeO_4)_3$		1.991	
$Ca_3Ti_2(GeO_4)_3$		1.924		$Mn_3Fe_2(GeO_4)_3$		2.044	
$Ca_3V_2(GeO_4)_3$		1.931		$Bi_{12}GeO_{20}^d$		2.5703	
$Ca_3Cr_2(GeO_4)_3$		1.924		$Bi_4(GeO_4)_3$		2.1077	

a[44], b[45], c[46], d[47]

References

1. R.M. Barrer, E.A.D. White, J. Chem. Soc. 1267 (1951)
2. A.S. Berger, T.I. Samsonova, I.A. Poroshina, Rus. J. Inorg. Chem. **17**, 1238 (1972) (in Russian)
3. C. Chuen-Lin, Acta Geol. Sinica **44**, 334 (1964)
4. I.O. Galuskina, B. Lazic, T. Armbruster et al., Amer. Miner. **94**, 1361 (2009)
5. N.V. Chukanov, V.V. Subbotin, I.V. Pekov et al., New Miner. **38**, 9 (2003) (in Russian)
6. N.V. Chukanov, B. Lazic, T. Armbruster, Amer. Miner. **97**, 1998 (2012)
7. W. Eitel, G. Trömel, Z. Krist. **73**, 67 (1930)
8. L.A. Pautov, A.A. Agakhanov, E. Sokolova, F.C. Hawthorne, Canad. Miner. **42**, 107 (2004)
9. R.M. Barrer, E.A.D. White, J. Chem. Soc. 1561 (1952)
10. J.T. Alfors, C.C. Stinson, R.A. Matthews, Amer. Miner. **50**, 314 (1965)
11. A.P. Khomyakov, L.I. Polezhaeva, S. Merlino et al., Proc. Miner. Soc. USSR **3**, 76 (1990) (in Russian)
12. N.A. Toropov, V.A. Vasil'eva, Crystallography **6**, 968 (1961) (in Russian)
13. N.A. Toropov, I.A. Bondar', Proc. Acad. Sci. USSR, Chem. 544 (1961) (in Russian)
14. R.M. Barrer, J.M. Baynham, J. Chem. Soc. 2882 (1956)
15. A. Kato, K. Nagasima, Geol. Surv. Jpn. 85 (1970)
16. N.A. Toropov, I.A. Bondar', Proc. Acad. Sci. USSR, Chem. **739**, 1372 (1961) (in Russian)
17. N.A. Toropov, F.Ya.Galakhov, S.F. Konovalova, Proc. Acad. Sci. USSR, Chem. **539**, 1365 (1961) (in Russian)
18. H.F.W. Taylor, J. Chem. Soc. 1253 (1949)
19. A. Pabst, C.O. Hutton, Amer. Miner. **36**, 60 (1951)
20. H. Sarp, I. Bertrand, E. McNear, Amer. Miner. **61**, 825 (1976)
21. C.M. Gramaccioli, W.L. Griffin, A. Mottana, Amer. Miner. **65**, 947 (1980)
22. L.B. Alexander, B.H. Feinter, Miner. Mag. **35**, 622 (1965)
23. L.Z. Reznitskiy, E.V. Sklyarov, Z.F. Ushchapovskaya, Proc. Miner. Soc. USSR **5**, 630 (1985) (in Russian)
24. A.S. Berger, L.T. Menzheres, N.P. Kotsupalo et al., Proc. Sib. Div. Acad. Sci. USSR **2**, 67 (1981) (in Russian)
25. E.J. Essene, D.R. Peacor, Amer. Miner. **72**, 157 (1987)
26. R.A. Kordyuk, N.V. Gul'ko, Dokl. Acad. Sci. USSR **142**, 639 (1962) (in Russian)
27. E.B. Gross, J.E. Wainwright, B.W. Evans, Amer. Miner. **50**, 1164 (1965)
28. B.R. Young, J.R. Hawkes, R.J. Merriman et al., Miner. Mag. **42**, 3 (1978)
29. V.E. Plyushchev, Dokl. Acad. Sci. USSR **124**, 642 (1959) (in Russian)
30. H. Hori, K. Nagashima, M. Yamada et al., Amer. Miner. **71**, 1022 (1986)
31. Z.D. Alekseeva, Rus. J. Inorg. Chem. **8**, 1426 (1963) (in Russian)
32. Z.D. Alekseeva, Rus. J. Inorg. Chem. **11**, 1171 (1966) (in Russian)
33. N.A. Toropov, R.G. Grebenschikov, Rus. J. Inorg. Chem. **7**, 337 (1962) (in Russian)
34. A.P. Khomyakov, E.I. Semenov, E.A. Pobedimskaya et al., Proc. Miner. Soc. USSR **6**, 80 (1991) (in Russian)
35. J. Gittins, M.G. Bown, D. Sturman, Canad. Miner. **14**, 120 (1976)
36. A. Pabst, Acta Cryst. **12**, 733 (1959)
37. S. Merlino, N. Perchiazzi, A.P. Khomyakhov et al., Eur. J. Miner. **2**, 177 (1990)
38. T.G. Sahama, J. Siivola, P. Rehtijärvi, Bull. Geol. Soc. Finland **43**, 1 (1973)
39. A.P. Khomyakhov, A.A. Voronkov, S.I. Lebedeva et al., Proc. Miner. Soc. USSR **103**(1), 110 (1974) (in Russian)
40. A.C. Roberts, M. Bonardi, R.C. Erd et al., Miner. Rec. **21**, 215 (1990)
41. H.G. Ansell, A.C. Roberts, A.G. Plant et al., Canad. Miner. **18**, 201 (1980)
42. P. Chi-Jui, T. Rung-Lung, Z. Zu-Rung, Sci. Sinica **11**, 977 (1962)
43. A.I. Leonov, V.S. Rudenko, E.K. Keler, Proc. Acad. Sci. USSR, Chem. 1925 (1961) (in Russian)

44. E.R. Shaw, J.F. Corwin, J.W. Edwards, J. Am. Chem. Soc. **80**, 1536 (1958)
45. C.R. Robbins, E.M. Levin, Amer. J. Sci. **257**, 63 (1959)
46. F. Dachille, R. Roy, Amer. J. Sci. **258**, 225 (1960)
47. M. Simon, F. Mersch, C. Kuper et al., Phys. Stat. Solidi a **159**, 559 (1997)

Chapter 7
Refractive Indices of Uranium Compounds

Because most of uranium oxides and their compounds contain water molecules, this table includes both anhydrous substances and crystallohydrates (Table 7.1).

© The Author(s) 2016   57
S.S. Batsanov et al., *Refractive Indices of Solids*,
SpringerBriefs in Applied Sciences and Technology,
DOI 10.1007/978-981-10-0797-2_7

Table 7.1 Refractive indices of the uranium oxides and their derivatives

Compounds	n_g	n_m	n_p
$NaUO_2(CH_3COO)_3{}^a$		1.501	
$NaUO_2PO_4 (H_2O)_4{}^b$		1.578	1.559
$Na_2CaUO_2(CO_3)_3(H_2O)_6$	1.540	1.520	
$Na_4UO_2(CO_3)_3{}^d$		1.645	1.531
$KUO_2VO_4{}^c$	2.10	1.98	1.71
$KUO_2AsO_4(H_2O)_4{}^f$		1.597	1.570
$K_2UO_2(NO_3)_4$	1.542	?	1.535
$K_2UO_2(SO_4)_2(H_2O)_2{}^g$	1.569	1.527	1.514
$K_3(UO_2)_2F_7(H_2O)_2{}^h$	1.502	1.459	1.448
$K_5(UO_2)_2F_9{}^i$	1.536	1.491	1.479
$Rb_2UO_2(NO_3)_4{}^j$	1.659	1.561	1.535
$Rb_2UO_2(SO_4)_2(H_2O)_2{}^g$	1.605	1.572	1.568
$NH_4UO_2(NO_3)_3{}^a$		1.640	1.472
$NH_4UO_2PO_4(H_2O)_3$		1.585	1.564
$(NH_4)_2UO_2(NO_3)_4{}^a$	1.670	1.562	1.544
$(NH_4)_2UO_2(NO_3)_4(H_2O)_2{}^a$	1.626	1.620	1.498
$(NH_4)_2UO_2Cl_4(H_2O)_2$	1.637	1.633	1.570
$(NMe_4)_2UO_2Cl_4^q$		1.526	1.516
$(NEt_4)_2UO_2Cl_4^q$	1.559	1.558	1.531
$NH_4(UO_2)_2(OH)_4F$		1.771	
$(NH_4)_2UO_2(SO_4)_2(H_2O)_2{}^a$	1.600	1.562	1.558
$(NH_4)_2UO_2(C_2O_4)_2(H_2O)_2{}^\alpha$	1.636	?	1.535
$(NH_4)_2UO_2(C_2O_4)_2(H_2O)_3{}^\alpha$	1.580	1.480	
$(NH_4)_3UO_2F_5$		1.486	
$MgCaUO_2(CO_3)_3(H_2O)_{12}$	1.540	1.510	1.465
$CaUO_2(CO_3)_2(H_2O)_5^c$	1.697	1.559	1.536
$Ca(UO_2)_2(PO_4)_2(H_2O)_7$		1.600	1.590
$Ca(UO_2)_2(PO_4)_2(H_2O)_8$	1.578	1.575	1.555
$Ca(UO_2)_2(PO_4)_2(H_2O)_{12}$	1.521	1.510	1.488
$Ca(UO_2)_2(VO_4)_2(H_2O)_4$	1.895	1.870	1.670
$Ba(UO_2)_2(PO_4)_2(H_2O)_6$	1.621	1.607	
$Ba(UO_2)_2(AsO_4)_2(H_2O)_8$	1.632	1.623	
$Pb(UO_2)_2(PO_4)_2(H_2O)_4{}^k$	1.752	1.749	1.739
$UO_2Cl_2H_2O^a$	1.700	1.676	1.672
$UO_2OHCl(H_2O)^a$	1.674	1.641	1.639
$UO_2CO_3^l$	1.795	1.716	1.70
$UO_2CO_3H_2O^m$	1.612	1.588	
$UO_2SO_3(H_2O)_3^n$	1.690	?	1.634
$UO_2SO_3(H_2O)_{4.5}{}^n$	1.641	?	1.559
$UO_2SO_4(H_2O)_3^o$	1.593	1.589	1.574
$UO_2C_2O_4(H_2O)_3^p$	1.634	1.486	1.476
$UO_2(CH_3COO)_2(H_2O)_2^a$	1.621	1.536	1.535
$UO_2(NO_3)_2(H_2O)_3^a$	1.610	1.586	1.504
$UO_2(NO_3)_2(H_2O)_6^a$	1.572	1.497	1.484
$UO_2Pb_2(CO_3)_3^r$	1.945	1.905	1.803
$(UO_2)_2SiO_4(H_2O)_2$	1.712	1.685	1.650
$\alpha\text{-}UO_3H_2O^t$	1.780	1.745	1.735
$\beta\text{-}UO_3H_2O^t$	1.880	1.855	1.850

(continued)

Table 7.1 (continued)

Compounds	n_g	n_m	n_p	Compounds	n_g	n_m	n_p
$Am_4(UO_2)_2(SO_4)_3 \cdot 5W^s$	1.583	1.582	1.560	$UO_3(H_2O)_2^u$	1.790	1.730	1.695
$Am_4UO_2(CO_3)_3$	1.625	1.62	1.60	$(UO_3)_3(H_2O)5$	1.830	1.822	1.735
$Cs_2UO_2Cl_4^a$	1.687	1.674	1.608	$(UO_3)_2(V_2O_5)_3(H_2O)_{15}$	2.057	1.879	1.817
$Cs_3UO_2F_5$	1.553	1.504	1.502	$UO_3(SiO_2)^y$	1.584	1.570	
$Cs_2(UO_2)_2(SO_4)_3^g$		1.643	1.615	$(UO_3)_5(SiO_2)_2(H_2O)_6^x$	1.712	1.685	1.650
$Cu(UO_2)_2(PO_4)_2 \cdot 12\ W$		1.592	1.582	$(UO_3)_6SO_3(H_2O)_{10}$		1.76	1.72
$Cu(UO_2)_2(AsO_4)_2 \cdot 8\ W$		1.647	1.629	$UO_3 \cdot (UO_2)_2CrO_4(H_2O)_4^\beta$	2.10	2.07	1.84
$CuUO_4 \cdot 2W^w$	1.80	1.792	1.765	$UO_4(H_2O)_2^z$	1.780	1.633	1.580
$Tl_4UO_2(CO_3)_3^y$	1.99	?	1.83	$UO_4(H_2O)_4^z$	1.712	1.579	1.533
$Mg(UO_2)_2(PO_4)_2 \cdot 8\ W$		1.574	1.559	$U(SO_4)_2(H_2O)_4^a$	1.659	1.620	1.588
$Mg_2UO_2(CO_3)_3 \cdot 18\ W$	1.500	1.490	1.455	$U(SO_4)_2(H_2O)_8^a$	1.580	1.544	1.530
				$U(C_2O_4)_2(H_2O)_6^a$	1.614	1.614	1.496

[a] [1], [b] [2], [c] [3], [d] [4], [e] [5], [f] [6], [g] [7], [h] [8], [i] [9], [j] [10], [k] [11], [l] [12], [m] [13], [n] [14], [o] [15], [p] [16], [q] [17], [r] [18], [s] [19], [t] [20], [u] [21], [v] [22], [w] [23], [x] [24], [y] [25], [z] [26], [α] [27], [β] [28]

References

1. E. Staritzky, A.L. Truitt, in *The actinide elements*, ed. by G.T. Seaborg, J.J. Katz (McGraw-Hill, New York, 1954)
2. A.A. Chernikov, O.V. Krutetskaya, N.I. Organova, Atomic Energy **3**(8), 135 (1957) (in Russian)
3. R.G. Coleman, D.R. Ross, R. Meyrowitz, Amer. Miner. **51**, 1567 (1966)
4. R.M. Douglass, Analyt. Chem. **28**, 1635 (1956)
5. M.J. de Abeledo, M.R. de Benyacar, R. Poljak, Analyt. Chem. **30**, 452 (1958)
6. M.E. Thompson, B. Ingrem, E.B. Gross, Amer. Miner. **41**, 82 (1956)
7. M. Ross, H.T. Evans, J. Inorg. Nucl. Chem. **15**, 338 (1960)
8. D.J. Walker, D.T. Cromer, E. Staritzky, Analyt. Chem. **28**, 1501 (1956)
9. E. Staritzky, D.T. Cromer, D.J. Walker, Analyt. Chem. **28**, 1355 (1956)
10. E. Staritzky, D.J. Walker, Analyt. Chem. **29**, 164 (1956)
11. R.V. Getseva, K.T. Savel'eva, *Determination of uranium minerals* (Moscow, Geology Press, 1956), p 213 (in Russian)
12. E. Staritzky, D.T. Cromer, Analyt. Chem. **28**, 1211 (1956)
13. R. Vochen, M. Deliens, Canad. Miner. **36**, 1077 (1998)
14. G.A. Polonnikova, K.F. Kudinova, Rus. J. Inorg. Chem. **6**, 1520 (1961) (in Russian)
15. V.L. Devshin, G.D. Sheremet'ev, J. Exp. Theor. Phys. **17**, 209 (1947) (in Russian)
16. E. Staritzky, D.T. Cromer, Analyt. Chem. **28**, 1353 (1956)
17. E. Staritzky, J. Singer, Acta Cryst. **5**, 536 (1952)
18. K. Walenta, Schweiz. Miner. Petrogr. Mitt. **56**, 167 (1976)
19. E. Staritzky, D.T. Cromer, D.J. Walker, Analyt. Chem. **28**, 1634 (1956)
20. L.A. Harris, A.J. Taylor, J. Amer. Ceram. Soc. **45**, 25 (1962)
21. J.W. Frondel, F. Cuttitta, Amer. Miner. **39**, 1018 (1954)
22. K. Walenta, Ns. Jb. Min. Monatsh. **6**, 259 (1983)
23. J.H. Milne, E.W. Nuffield, Amer. Miner. **36**, 394 (1951)
24. D.H. Gorman, Amer. Miner. **37**, 386 (1952)
25. T.N. Burlakova, Bull. Leningrad Univ. **178**, 157 (1954) (in Russian)
26. E.D. Ruchkin, S.A. Durasova, Proc. Sib. Div. Acad. Sci. USSS, **7**(2), 62 (1964) (in Russian)
27. I.I. Chernyaev, V.A. Golovnya, R.N. Schelokov, Rus. J. Inorg. Chem. **2**, 1763 (1957) (in Russian)
28. R.M. Douglass, E. Staritzky, Analyt. Chem. **29**, 314 (1954)

Chapter 8
Refractive Indices of Oxygen-Containing Salts

This section lists crystalline compounds containing such anions as $[E_mO_n]^{-q}$ where E = C, N, P, As, S and Se (Tables 8.1, 8.2, 8.3).

Table 8.1 Refractive indices of carbonates and nitrates

Compounds	n_g	n_m	n_p	Compounds	n_g	n_m	n_p
Li_2CO_3	1.572	1.567	1.428	$Na_3Mg(CO_3)_2Cl$		1.514	
$LiNaCO_3$		1.538	1.406	$Na_3Mg(CO_3)_2Br$		1.515	
Na_2CO_3	1.546	1.535	1.415	$Na_3Ca_2(CO_3)_3F$[a]	1.569	1.562	1.472
$NaHCO_3$	1.586	1.500	1.380	$Na_3Ce_2(CO_3)_4F$[b]		1.728	1.542
K_2CO_3	1.541	1.531	1.426	$Na_6Mg_2(CO_3)_4SO_4$		1.510	
$KHCO_3$	1.573	1.482	1.380	$Na_6Mn_2(CO_3)_4SO_4$[c]		1.544	
NH_4HCO_3	1.555	1.536	1.423	$Na_6Mg_2(CO_3)_4CrO_4$		1.555	
$MgCO_3$		1.700	1.509	$K_2Mg(CO_3)_2$		1.597	1.470
$MgCa(CO_3)_2$		1.679	1.502	$K_2Ca(CO_3)_2$		1.530	1.48
$CaCO_3$[d]		1.658	1.486	$Cu_2CO_3(OH)_2$	1.909	1.875	1.655
$CaCO_3$[e]		1.645	1.550	$Cu_3(CO_3)_2(OH)_2$	1.836	1.754	1.730
$CaCO_3$[f]	1.685	1.681	1.530	$Ca_2CO_3F_2$[g]	1.593	1.590	1.525
$CaBa(CO_3)_2$		1.673	1.525	$Ca_3CO_3SiO_4$	1.652	1.635	1.617
$CaMn(CO_3)_2$		1.741	1.536	$\alpha\text{-}Ca_5CO_3(SiO_4)_2$	1.680	?	1.665
$CaFe(CO_3)_2$		1.765	1.555	$\beta\text{-}Ca_5CO_3(SiO_4)_2$	1.679	1.674	1.640
$SrCO_3$	1.668	1.667	1.520	$BaCa_2(CO_3)_2F_2$[h]	1.614	1.612	1.500
$BaCO_3$	1.677	1.676	1.529	$BaCe(CO_3)_2F$[i]		1.765	1.603

(continued)

© The Author(s) 2016
S.S. Batsanov et al., *Refractive Indices of Solids*,
SpringerBriefs in Applied Sciences and Technology,
DOI 10.1007/978-981-10-0797-2_8

Table 8.1 (continued)

Compounds	n_g	n_m	n_p	Compounds	n_g	n_m	n_p
$ZnCO_3$		1.848	1.621	$Ba_2Ce(CO_3)_3F^j$	1.728	1.724	1.584
$PbCO_3$	2.078	2.076	1.804	$Zn_5CO_3(OH)_6$	1.750	1.736	1.640
$MnCO_3$		1.816	1.597	YCO_3OH		1.596	
$FeCO_3$		1.875	1.633	$Ce_2(CO_3)_3^k$	1.603	1.575	1.528
$CoCO_3$		1.855	1.600	$NdCO_3OH^l$	1.780	?	1.698
$Ce_2(CO_3)_3^k$	1.603	1.575	1.528	$DyCO_3F$		1.566	
$(BiO)_2CO_3$		2.13	1.94	$DyCO_3OH$		1.606	
$Ca(BiO)_2(CO_3)_2$	2.13	1.99		$ErCO_3OH$		1.604	
$NaCa(CO_3)_2^m$	1.570	1.555	1.531	$Pb_2CO_3Cl_2$	2.145	2.118	
$Na_2Ca(CO_3)_2$		1.547	1.504	$Pb_3(CO_3)_2(OH)_2$		2.09	1.94
$Na_2Mg(CO_3)_2$		1.594	1.540	$Pb_4(CO_3)_2SO_4(OH)_2$	2.01	2.00	1.87
$Na_2Cu(CO_3)_2^n$		1.571		$Fe_2CO_3(OH)_6^p$	1.780	1.770	1.673
$NaYCO_3F_2^o$	1.622	1.543	1.457	$Ni_2CO_3(OH)_2^q$	1.78	~1.78	1.67
$NaAlCO_3(OH)_2$	1.596	1.542	1.466				
$LiNO_3$		1.735	1.435	$Sr(NO_3)_2$		1.590	
$LiCs(NO_3)_2^r$	1.692	?	1.500	$Ba(NO_3)_2$		1.569	
$NaNO_3$		1.587	1.336	$Pb(NO_3)_2$		1.782	
KNO_3	1.5064	1.5056	1.3346	$Cu(NO_3)_2^k$	1.734	1.719	1.709
$RbNO_3$	1.524	1.520	1.51	$Cu_2NO_3(OH)_3$	1.722	1.713	1.703
o-$NH_4NO_3^s$	1.637	1.611	1.413	$La(NO_3)_3$		1.598	
h-$NH_4NO_3^s$	1.623	1.493		$Ce(NO_3)_3$		1.575	
t-$NH_4NO_3^s$	1.585	1.509		$Pr(NO_3)_3$		1.59	
$CsNO_3$	1.560	1.558		$Nd(NO_3)_3$		1.605	
$AgNO_3$	1.788	1.744	1.729	$Bi(NO_3)_3^k$	1.680	?	1.48
$TlNO_3$	1.869	1.862	1.817				

[a][1], [b][2], [c][3], [d]calcite, [e]faterite, [f]aragonite, [g][4], [h][5], [i][6], [j][7], [k][8], [l][9], [m][10], [n][11], [o][12], [p][13], [q][14], [r][15], [s]o orthorhombic, t tetragonal, h hexagonal polymorphs

Table 8.2 Refractive indices of phosphates and arsenates

Compounds	n_g	n_m	n_p	Compounds	n_g	n_m	n_p
$HNa_3Mg(PO_4)_2^a$	1.568	1.542	1.536	$Ca_3(PO_4)_2$		1.629	1.626
$HK_3Ca(PO_4)_2^a$	1.575	1.550	1.538	$Ca_4O(PO_4)_2$	1.656	1.651	1.650
$H_2KFe(PO_4)_2$	1.680	1.665	1.631	$Ca_5(PO_4)_3F$		1.632	1.630
$H_2(NH_4)Fe(PO_4)_2$	1.67	?	1.66	$Ca_5(PO_4)_3OH$		1.651	1.644
H_3PO_4	1.505	1.504	1.455	$Ca_5(PO_4)_3Cl$		1.6684	1.6675
$LiMnPO_4$	1.673	1.668	1.663	$Ca_5(PO_4)_2SiO_4$	1.640	1.636	1.632

(continued)

Table 8.2 (continued)

Compounds	n_g	n_m	n_p	Compounds	n_g	n_m	n_p
LiAlOHPO$_4$	1.6480	1.6290	1.6211	Ca$_7$(PO$_4$)$_2$(SiO$_4$)$_2^b$	1.661	?	1.652
Li$_2$NaPO$_4^c$	1.541	1.540	1.533	α-SrHPO$_4$	1.624	1.608	1.593
Li$_3$PO$_4^d$	1.562	1.555	1.550	Sr$_5$(PO$_4$)$_3$F		1.621	1.619
NaH$_2$PO$_4$	1.517	1.507	1.481	Sr$_5$(PO$_4$)$_3$Cl		1.655	1.650
NaBePO$_4$	1.561	1.558	1.552	BaHPO$_4$	1.635	1.63	1.617
NaCaPO$_4^f$	1.616	1.610	1.607	BaAl$_2$(PO$_4$)$_2$(OH)$_2^e$	1.710	1.693	1.672
NaMnPO$_4$	1.684	1.674	1.671	Ba$_3$(PO$_4$)$_2$	1.645	?	1.514
NaFePO$_4^g$	1.698	1.695	1.676	Ba$_5$(PO$_4$)$_3$F		1.669	1.665
Na$_2$HPO$_4$	1.525	1.499	1.483	Ba$_5$(PO$_4$)$_3$Cl		1.701	1.699
Na$_2$BeZr$_2$(PO$_4$)$_4^i$	1.630	1.618		ZnFe$_2$(PO$_4$)$_2$(OH)$_2^h$		1.795	1.755
Na$_2$Fe$_2$Al(PO$_4$)$_3^k$	1.696	1.691	1.688	ZnFe$_4$(PO$_4$)$_3$(OH)$_5^j$	1.767	1.764	1.756
Na$_3$PO$_4$	1.508	1.499	1.493	Zn$_2$PO$_4$OH	1.713	1.705	1.660
Na$_3$SrPO$_4$CO$_3^l$	1.565	1.564	1.520	Zn$_4$O(PO$_4$)$_2$	1.660	1.656	1.630
Na$_3$MnPO$_4$CO$_3^n$	1.585	1.563	1.521	ScPO$_4^m$		1.790	1.786
Na$_5$Ca$_4$(PO$_4$)$_4$F^o	1.590	1.571		YPO$_4$	1.831	1.652	
KH$_2$PO$_4$		1.5095	1.4684	CePO$_4$	1.843	1.794	1.793
KTiOPO$_4$	1.8724	1.7780	1.7683	TbPO$_4$	1.866	1.677	
KZr$_2$(PO$_4$)$_3$ ^{p}	1.682	1.656		DyPO$_4$	1.861	1.682	
KGeOPO$_4$	1.6732	1.6636	1.6585	HoPO$_4$	1.861	1.678	
RbH$_2$PO$_4$		1.5054	1.4764	ErPO$_4$	1.852	1.676	
RbTiOPO$_4$	1.8947	1.8057	1.7943	TmPO$_4$	1.849	1.678	
NH$_4$H$_2$PO$_4$		1.5246	1.4792	YbPO$_4$	1.853	1.676	
(NH$_4$)$_2$HPO$_4$	1.582	1.570	1.468	LuPO$_4$	1.855	1.675	
CuZnPO$_4$OH^q	1.715	1.705	1.660	cr-BPO$_4$	1.6013	1.5947	
CuFe$_2$(PO$_4$)$_2$(OH)$_2^s$	1.945	1.848	1.843	qu-BPO$_4^r$		1.642	
Cu$_2$PO$_4$OH	1.787	1.743	1.701	qu-AlPO$_4$	1.530	1.523	
Cu$_3$PO$_4$(OH)$_3$	1.825	1.820	1.762	cr-AlPO$_4^r$		1.465	
Cu$_5$(PO$_4$)$_2$(OH)$_4^s$	1.867	1.833	1.782	Al$_2$PO$_4$(OH)$_3$	1.588	?	1.575
Ag$_2$HPO$_4$		1.8036	1.7983	qu-GaPO$_4^r$		1.603	
Be$_2$Ca(PO$_4$)$_2^t$	1.604	1.601	1.595	cr-GaPO$_4^r$		1.560	
Mg$_2$PO$_4$F	1.5824	1.5719	1.5678	PbHPO$_4$	1.885	1.822	
Mg$_2$PO$_4$OH	1.552	1.534	1.533	Pb$_3$(PO$_4$)$_2$		1.9702	1.9364
Mg$_3$(PO$_4$)$_2^u$	1.559	1.544	1.540	Pb$_5$(PO$_4$)$_3$Cl		2.058	2.048
Mg$_7$(PO$_4$)$_2$(OH)$_8^w$	1.609	1.607	1.594	Bi$_2$OPO$_4$OH^v	2.09	2.06	2.05
CaHPO$_4$	1.640	1.615	1.587	Bi$_3$O(PO$_4$)$_2$OH^x	2.13	?	2.06

(continued)

Table 8.2 (continued)

Compounds	n_g	n_m	n_p	Compounds	n_g	n_m	n_p
Ca(H$_2$PO$_4$)$_2^y$	1.601	1.580	1.547	qu-MnPO$_4$	1.532	1.525	
Ca$_2$PO$_4$Cl	1.670	1.663	1.650	cr-MnPO$_4^r$		1.482	
CaZn$_2$(PO$_4$)$_2$(H$_2$O)$_2^z$	1.603	1.588	1.587	Mn$_7$(PO$_4$)$_2$(OH)$_8^\alpha$	1.738	1.738	1.730
Ca$_2$Al(PO$_4$)$_2$OH$^\beta$	1.696	1.671	1.662	FePO$_4$	1.77	1.75	
Ca$_2$Pb$_3$(PO$_4$)$_3$Cl$^\gamma$		1.935	1.928	PuPO$_4^\delta$	1.905	1.86	1.855
H$_2$NaZn$_3$ (AsO$_4$)$_3^\varepsilon$	1.778	1.753	1.745	SrHAsO$_4$	1.67	?	1.635
H$_4$Ca$_2$AsO$_4$BO$_4$	1.663	1.662		Ba$_5$(AsO$_4$)$_3$Cl$^\eta$	1.884	1.880	
H$_4$Pb (AsO$_4$)$_2$		1.82	1.74	ZnFe$_2$(AsO$_4$)$_2$(OH)$_2^\theta$		1.94	
NaAlAsO$_4$F	1.685	1.673	1.634	Zn$_2$AsO$_4$OH	1.761	1.742	1.722
NaFe(AsO$_4$)F$^\kappa$	1.798	1.772	1.748	YAsO$_4^\lambda$	1.879	1.783	
NaCu$_4$(AsO$_4$)$_3^\mu$	1.96	1.92	1.76	LaAl$_3$(AsO$_4$)$_2$(OH)$_6^\nu$	1.750	1.740	
KH$_2$AsO$_4$		1.5674	1.5179	BAsO$_4$	1.690	1.681	
NH$_4$H$_2$AsO$_4$		1.5766	1.5217	AlAsO$_4^\xi$	1.608	1.591	
CuZnAsO$_4$OH$^\pi$	1.788	1.784	1.736	PbHAsO$_4$	1.9765	1.9097	1.8903
CuAlOAsO$_4^\rho$	1.722	1.718	1.672	PbFe$_3$H (AsO$_4$)$_2$(OH)$_6^\sigma$		1.975	1.955
CuPbAsO$_4$OH	2.08	2.06	2.03	Pb$_3$(AsO$_4$)$_2$Cl		2.147	2.128
CuFeOAsO$_4^\tau$	1.910	1.865	1.830	Pb$_5$ (AsO$_4$)$_3$Cl	2.15	2.12	
Cu$_2$AsO$_4$OH	1.85	1.80	1.76	β-BiAsO$_4^\upsilon$		2.20	
Cu$_3$(AsO$_4$)$_2^\varphi$	1.955	1.915	1.885	BiFeOAsO$_4$OH$^\chi$	2.12	2.09	2.02
Cu$_3$AsO$_4$ (OH)$_3^\psi$	1.896	1.874	1.756	Bi$_3$O(AsO$_4$)$_2$OH$^\omega$	2.195	2.160	2.130
Cu$_5$ (AsO$_4$)$_2$(OH)$_4$	1.88	1.86	1.82	Mn$_2$AsO$_4$OH	1.809	1.807	1.793
BeCaAsO$_4$OH$^\Lambda$	1.694	1.681	1.659	Mn$_5$ AsO$_4$)$_2$(OH)$_4$	1.816	1.810	1.787
CaHAsO$_4$	1.653	1.650	1.635	Mn$_7$ AsO$_4$)$_2$(OH)$_8$	1.774	1.772	1.755
CaMn$_2$(AsO$_4$)$_2^B$	1.790	1.785	1.784	Mn$_4$AsO$_4$(OH)$_5$	1.761	1.751	1.750
CaCoAsO$_4$OH^C	1.802	?	1.777	FeAsO$_4$		1.78	
Ca$_5$(AsO$_4$)$_3$OH^D		1.687	1.684				

^{a}[16], ^{b}[17], ^{c}[18], ^{d}[19], ^{e}[20], ^{f}[21], ^{g}[22], ^{h}[23], ^{i}[24], ^{j}[25], ^{k}[26], ^{l}[27], ^{m}[28], ^{n}[29], ^{o}[30], ^{p}[31], ^{q}[32], ^{r}[33], cr cristobalite, qu quartz-type structure, ^{s}[34], ^{t}[35], ^{u}[36], ^{v}[37], ^{w}[38], ^{x}[39], ^{y}[40], ^{z}[41], $^\alpha$[42], $^\beta$[43], $^\gamma$[44], $^\delta$[45], $^\varepsilon$[46], $^\eta$[47], $^\theta$[48], $^\kappa$[49], $^\lambda$[50], $^\mu$[51], $^\nu$[52], $^\xi$[53], $^\pi$[54], $^\rho$[55], $^\sigma$[56], $^\tau$[57], $^\upsilon$[58], $^\varphi$[59], $^\chi$[60], $^\psi$[61], $^\omega$[62], $^\Lambda$[63], ^{B}[64], ^{C}[65], ^{D}[66]

Table 8.3 Refractive indices of sulphates, selenates and perchlorates

Compounds	n_g	n_m	n_p	Compounds	n_g	n_m	n_p
Li_2SO_4		1.465		$NH_4Fe_3(SO_4)_2(OH)_6$		1.830	1.745
$LiNaSO_4$	1.495	1.490		$(NH_4)_2Ca_2(SO_4)_3$		1.532	
$LiKSO_4$		1.4721	1.4715	$(NH_4)_2Ti(SO_4)_3$ [a]	1.754	?	1.684
$LiNH_4SO_4$		1.437		α-$(NH_4)_2TiO(SO_4)_2$ [b]	1.707	?	1.600
$NaFe_3(SO_4)_2(OH)_6$		1.832	1.750	β-$(NH_4)_2TiO(SO_4)_2$ [b]		1.629	
Na_2SO_4	1.484	1.477	1.471	$(NH_4)_2Pb(SO_4)_2$ [c]		1.721	1.642
$Na_2Mg(SO_4)_2$ [e]	1.650	?	1.450	$(NH_4)_3H(SO_4)_2$	1.528	1.502	1.496
$Na_2Ca(SO_4)_2$	1.536	1.535	1.515	$(NH_4)_3Nb(SO_4)_4$ [d]	1.645	?	1.516
$Na_3H(SO_4)_2$ [f]	1.479	?	1.459	$(NH_4)_4SO_4\ NO_3$	1.552	1.531	1.521
Na_3SO_4F [g]	1.442	1.439		$(NH_4)_5\ SO_4(NO_3)_3$		1.540	1.488
Na_3SO_4OH [h]	1.477	?	1.471	Cs_2SO_4	1.5662	1.5644	1.5598
$Na_3Ca_2(SO_4)_3OH$ [i]		1.570	1.564	$CuSO_4$	1.739	1.733	1.724
$Na_4Bi(SO_4)_3Cl$ [j]	1.60	1.59		$CuPbSO_4(OH)_2$	1.8593	1.8380	1.8090
$Na_6Mg(SO_4)_4$	1.4893	1.4876	1.4855	Cu_2OSO_4	1.880	1.820	1.715
$Na_6(SO_4)_2CO_3$	1.492	1.490	1.450	$Cu_3SO_4(OH)_4$	1.789	1.738	1.726
$Na_6Fe_2SO_4(CO_3)_4$ [k]		1.550		$Cu_4SO_4(OH)_6$	1.800	1.771	1.728
K_2SO_4	1.4973	1.4947	1.4935	Ag_2SO_4	1.7852	1.7748	1.7583
$KHSO_4$	1.494	1.452	1.438	$AgFe_3(SO_4)_2(OH)_6$		1.882	1.785
$K_3H(SO_4)_2$	1.526	1.490	1.479	$MgSO_4$		~1.57	
$KAl(SO_4)_2$ [l]		1.546	1.533	$CaSO_4$	1.6136	1.5754	
$KFe(SO_4)_2$ [n]	1.698	1.684	1.593	$CaTi(SO_4)_3$		1.474	
$KCu_3(SO_4)_2Cl$ [o]	1.759	1.718	1.695	$Ca_4Al_6O_{12}SO_4$ [m]		1.568	
$K_2AlSO_4F_3$ [p]		1.445		$SrSO_4$	1.6305	1.6232	1.8600
$K_2Mg_2(SO_4)_3$		1.5347		$BaSO_4$	1.6484	1.6373	1.5698

(continued)

Table 8.3 (continued)

Compounds	n_g	n_m	n_p	Compounds	n_g	n_m	n_p
$K_2Ca_2(SO_4)_3^q$		1.527		$ZnSO_4$	1.670	1.669	1.6215
$K_2Cu_2O(SO_4)_2^r$	1.695	1.583	1.533	Hg_2SO_4	2.21	1.810	1.796
$K_2Al_2(SO_4)_4$		1.545		$HgSO_4$	>1.810	1.810	
$K_2Mn_2(SO_4)_3$		1.572	1.572	$Pr_2O_2SO_4^s$	1.921	1.917	1.826
$KAl_3(SO_4)_2(OH)_6$		1.592	1.572	$Al_2(SO_4)_3$		1.47	
$KFe_3(SO_4)_2(OH)_6$		1.820	1.715	Tl_2SO_4	1.8853	1.8671	
$K_2Cu_3O(SO_4)_3^u$	1.633	1.594	1.577	$\alpha\text{-}TlHSO_4^t$	1.702	1.690	1.682
$K_3H(SO_4)_2$	1.5259	1.4899	1.4793	$\beta\text{-}TlHSO_4^t$	1.684	1.679	1.653
$K_3Cu_3AlO_2(SO_4)_4^v$	1.641	1.548	1.542	$Tl_2H_4(SO_4)_3^t$		1.619	
$K_3Cu_3FeO_2(SO_4)_4^w$	1.680	1.550	1.549	$Tl_3H(SO_4)_2^t$		1.760	1.730
$K_8Pu(SO_4)_7^x$	1.605	?	1.576	$PbSO_4$	1.8947	1.8832	1.6363
Rb_2SO_4	1.5144	1.5133	1.5131	$PbFe_6(SO_4)_4(OH)_{12}$		1.875	1.786
$RbHSO_4$		1.473		Pb_2OSO_4	2.036	2.007	1.928
$RbFe_3(SO_4)_2(OH)_6$		1.805	1.720	$Pb_2SO_4F_2^y$	1.897	1.873	1.872
$(NH_4)_2SO_4$	1.5330	1.5230	1.5209	$Pb_2SO_4(OH)_2$		1.93	
NH_4HSO_4	1.510	1.473	1.463	$Pb_2SO_4SeO_4^z$	1.983	1.966	1.936
$NH_4Pr(SO_4)_2$	1.623	?	1.598	$Pb_4SO_4(CO_3)_2(OH)_2^\alpha$	2.01	2.00	1.87
$NH_4Nd(SO_4)_2$	1.623	?	1.599	$Pb_6SO_4F_{10}$	1.873	1.865	1.864
$NH_4Sm(SO_4)_2$	1.626	?	1.600	$VOSO_4^\gamma$	1.845	1.778	1.731
$NH_4Eu(SO_4)_2$	1.629	1.621	1.602	$h\text{-}Fe_2(SO_4)_3$		1.756	1.746
$NH_4Gd(SO_4)_2$	1.629	1.620	1.604	$o\text{-}Fe_2(SO_4)_3$		1.814	1.802
$NH_4Al_3(SO_4)_2(OH)_6^\beta$	1.602	1.590		$FeSO_4OH$	1.918	1.804	1.783
$(NH_4)_2Mg_2(SO_4)_3^\delta$		1.550					
K_2SeO_4	1.5446	1.5390	1.5352	$NH_4La(SeO_4)_2$	1.684	1.676	1.658

(continued)

Table 8.3 (continued)

Compounds	n_g	n_m	n_p	Compounds	n_g	n_m	n_p
Rb_2SeO_4	1.5582	1.5537	1.5515	$NH_4Pr(SeO_4)_2$	1.700	1.695	1.677
$(NH_4)_2SeO_4$	1.5846	1.5630	1.5607	$NH_4Nd(SeO_4)_2$	1.703	1.696	1.680
Cs_2SeO_4	1.6003	1.5999	1.5989	$NH_4Sm(SeO_4)_2$	1.704	1.696	1.680
Tl_2SeO_4	1.9640	1.9592	1.9493	$NH_4Eu(SeO_4)_2$	1.705	?	1.683
$CuSeO_4^\varepsilon$	1.728	?	1.708	$NH_4Gd(SeO_4)_2$	1.705	1.696	1.681
$BeSeO_4^\eta$		1.542		$(NH_4)_3Er_2(SeO_4)_4NO_3$	1.660	1.633	1.580
$MgSeO_4^\theta$		1.585		$(NH_4)_3Yb_2(SeO_4)_4NO_3$	1.666	1.637	1.581
$PbSeO_4^\kappa$	1.98	?	1.96	NH_4ClO_4	1.4881	1.4833	1.4818
$NaClO_4$	1.4730	1.4617	1.4606	$CsClO_4$	1.4804	1.4788	1.4752
$KClO_4$	1.4769	1.4737	1.4731	$TlClO_4$	1.6541	1.6445	1.6427
$RbClO_4$	1.4731	1.4701	1.4692	$Cd(ClO_4)_2$		1.510	

[a][67], [b][68], [c][69], [d][70], [e][71], [f][72], [g][73], [h][74], [i][75], [j][76], [k][77], [l][78], [m][79], [n][80], [o][81], [p][82], [q][83], [r][84], [s][85], [t][86], [u][87], [v][88], [w][89], [x][90], [y][91], [z][92], [α][93], [β][94], [γ][95], [δ][96], [ε][97], [η][98], [θ][99], [κ][100]

References

1. A.M. McDonald, G.Y. Chao, R.A. Ramik, Canad. Miner. **29**, 107 (1991)
2. J.D. Grice, G.Y. Chao, Amer. Miner. **82**, 1255 (1997)
3. A.P. Khomyakov, A.Y. Bakhchisaraitsev, A.V. Martynova et al., Proc. Miner. Soc. USSR **119**, 46 (1990) (in Russian)
4. G. Hentschel, V. Leufer, E.N. Tillmanns, Ns. Jb. Min. Monatsh. **7**, 325 (1978)
5. I.V. Pekov, N.V. Zubkova, N.V. Chukanov et al., Miner. Records **39**, 137 (2008)
6. E.I. Semenov, P.S. Chzhan, Sci. Sinica **10**, 1007 (1961)
7. A.N. Zaitsev, V.N. Yakovenchuk, G.Y. Chao et al., Eur. J. Miner. **8**, 1327 (1996)
8. O.M. Ansheles, T.N. Burakova, *Crystallooptics as foundation of microchemical analysis* (Leningrad University Press, 1948) (in Russian)
9. R. Miyawaki, S. Matsubara, K. Yokoyama et al., Amer. Miner. **85**, 1076 (2000)
10. J.J. Faney, Amer. Miner. **24**, 514 (1930)
11. J. Schlüter, D. Pohl, Amer. Miner. **91**, 1204 (2006)
12. J.D. Grice, G.Y. Chao, Canad. Miner. **35**, 743 (1997)
13. I.V. Pekov, N. Perchiazzi, S. Merlino et al., Eur. J. Miner. **19**, 891 (2007)
14. E.H. Nickel, L.G. Berry, Miner. Mag. **19**, 315 (1981)
15. K.A. Bol'shakov, B.I. Pokrovsky, V.E. Plyuschev, Rus. J. Inorg. Chem. **6**, 2120 (1961) (in Russian)
16. A. Frazier, J. Smith, J. Lehr, W. Brown, Inorg. Chem. **1**, 949 (1962)
17. G. Nagelschmidt, J. Chem. Soc. 865 (1937)
18. G.Y. Chao, T.S. Ercit, Canad. Miner. **29**, 565 (1991)
19. T.Y. Tien, F.A. Hummel, J. Amer. Ceram. Soc. **44**, 206 (1961)
20. E.P. Meagher, M.E. Coates, Canad. Miner. **12**, 135 (1973)
21. E. Olsen, J. Erlichman, T.E. Bunch et al., Amer. Miner. **62**, 362 (1977)
22. B.D. Sturman, J.A. Mandarino, Canad. Miner. **15**, 396 (1977)
23. N.V. Chukanov, I.V. Pekov, Sh. Mekkel' et al., Proc. Rus. Miner. Soc. **6**, 13 (2006) (in Russian)
24. P.B. Moore, T. Araki, I.M. Steele et al., Amer. Miner. **68**, 1022 (1983)
25. P. Elliott, U. Kolitsch, G. Giester et al., Miner. Mag. **73**, 131 (2009)
26. P.B. Moore, J. Ito, Miner. Rec. **4**, 131 (1973)
27. A.P. Khomyakov, L.I. Polezhaeva, E.V. Sokolova, Proc. Miner. Soc. USSR **1**, 107 (1994) (in Russian)
28. H.-J. Bernhardt, F. Walter, K. Ettinger et al., Amer. Miner. **83**, 625 (1998)
29. A.P. Khomyakov, E.I. Semenov, M.E. Kazakova et al., Proc. Miner. Soc. USSR **108**, 56 (1979) (in Russian)
30. A.P. Khomyakov, G.N. Nechelyustov, G.I. Dorokhova, Proc. Miner. Soc. USSR **112**, 479 (1983) (in Russian)
31. E.E. Foord, M.E. Brownfield, F.E. Lichte et al., Canad. Miner. **32**, 839 (1994)
32. R.S. Braithwaite, R.C. Pritchard, W.H. Paar et al., Miner. Mag. **69**, 145 (2005)
33. F. Dachille, R. Roy, Z. Krist. **111**, 462 (1959)
34. N.H.W. Sieber, E. Tillmanns, O. Medenbach, Amer. Miner. **72**, 404 (1987)
35. M.E. Mrose, Amer. Miner. **37**, 931 (1952)
36. E.R. Fresne, S.K. Roy, Geochim. Cosmochim. Acta **24**, 198 (1961)
37. T. Ridkošil, J. Sejkora, V. Šrein et al., Ns. Jb. Min. Monatsh. **1**, 97 (1996)
38. C. Chopin, G. Ferraris, M. Prencipe et al., Eur. J. Miner. **13**, 319 (2001)
39. W. Krause, K. Belendorff, H.-J. Bernhardt, Ns. Jb. Min. Monatsh. **4**, 487 (1993)
40. R.C.L. Mooney, M.A. Aia, Chem. Rev. **61**, 433 (1961)
41. B.D. Sturman, R.C. Rouse, P.J. Dunn, Amer. Miner. **66**, 843 (1981)
42. A. Pring, U. Kolitsch, W. Birch, Canad. Miner. **43**, 1401 (2005)
43. C. Chopin, F. Brunet, W. Gerbert et al., Schweiz. Miner. Petrogr. Mitt. **73**, 1 (1993)
44. A.R. Kampf, I.M. Steele, R.A. Jenkins, Amer. Miner. **91**, 1909 (2006)

45. C.W. Bjorklund, J. Am. Chem. Soc. **79**, 6347 (1957)
46. P. Keller, H. Hess, Ns. Jb. Miner. Monatsh. **4**, 155 (1981)
47. P.J. Dunn, R.C. Rouse, Canad. Miner. **16**, 601 (1978)
48. J. Schlüter, K.-H. Klaska, K. Friese et al., Ns. Jb. Min. Monatsh. **4**, 558 (1998)
49. E.E. Foord, P.F. Hlava, J.J. Fitzpatrick et al., Ns. Jb. Min. Monatsh. **8**, 36 (1991)
50. B.A. Goldin, N.P. Yushkin, M.V. Fishman, Proc. Miner. Soc. USSR **96**, 669 (1967) (in Russian)
51. W. Krause, H.-J. Bernhardt, H. Effenberger et al., Eur. J. Miner. **14**, 115 (2002)
52. S.J. Mills, P.M. Kartashov, A.R. Kampf et al., Eur. J. Miner. **2**(2), 613 (2010)
53. T.F. Semenova, L.P. Vergasova, S.K. Filatov et al., Dokl. Rus. Acad. Sci. **338**, 501 (1994) (in Russian)
54. N.V. Chukanov, D.Y. Puscharovsky, N.V. Zubkova et al., Dokl. Earth Sci. **415**, 841 (2007)
55. S.V. Krivovichev, A.V. Molchanov, S.K. Filatov, Crystallogr. Rep. **45**, 723 (2000)
56. W.D. Birch, A. Pring, B.M. Gatehouse, Amer. Miner. **77**, 656 (1992)
57. S.J. Mills, A.R. Kampf, G. Poirier et al., Miner. Petrol. **99**, 113 (2010)
58. J. Sejkora, T. Ridkošil, Ns. Jb. Min. Monatsh. **4**, 179 (1994)
59. P. Keller, W.H. Paar, P.J. Dunn, Aufschluss **32**, 437 (1981)
60. W. Krause, H.-J. Bernhardt, C. McCammon et al., Amer. Miner. **87**, 726 (2002)
61. H. Sarp, R. Cerny, Eur. J. Miner. **11**, 549 (1999)
62. D. Bedlivy, K. Mereiter, Amer Miner. **67**, 833 (1982)
63. S. Hansen, L. Fälth, O.V. Petersen et al., Ns. Jb. Miner. Monatsh. **6**, 257 (1984)
64. S. Graeser, H. Schwander, B. Suhner, Schweiz. Miner. Petrogr. Mitt. **64**, 1 (1984)
65. E.H. Nickel, W.D. Birch, Austral. Miner. **3**, 53 (1988)
66. P.J. Dunn, D.R. Peacor, N. Newberry, Amer. Miner. **65**, 1143 (1980)
67. Y.G. Goroshchenko, Dokl. Acad. Sci. USSR **109**, 532 (1956) (in Russian)
68. D.L. Motov, Rus. J. Inorg. Chem. **2**(2661), 2797 (1957) (in Russian)
69. C.K. Møller, Acta Chem. Scand. **8**, 81 (1954)
70. Y.G. Goroshchenko, Rus. J. Inorg. Chem. **1**, 909 (1956) (in Russian)
71. I.G. Druzhinin, Dokl. Acad. Sci. USSR **23**, 914 (1939) (in Russian)
72. Ya.E. Vil'nyansky, Z.L. Persiu, Rus. J. Gen. Chem. **16**, 1997 (1946) (in Russian)
73. A. Pabst, W.N. Sharp, Amer. Miner. **58**, 116 (1973)
74. L.S. Itkina, Rus. J. Inorg. Chem. **7**, 184 (1962) (in Russian)
75. G. Cavarretta, Miner. Mag. **44**, 269 (1981)
76. F. Demartin, C.M. Gramaccioli, I. Campostrini et al., Amer. Miner. **95**, 382 (2010)
77. A.P. Khomyakov, Y.A. Malinovsky, S.M. Sandomirskaya, Proc. Miner. Soc. USSR **110**, 600 (1981) (in Russian)
78. M.N. Murashko, I.V. Pekov, S.V. Krivovichev et al., Proc. Rus. Miner. Soc. **4**, 36 (2012) (in Russian)
79. S. Gross, Amer. Miner. **72**, 226 (1987)
80. C.O. Hutton, Amer. Miner. **44**, 1105 (1959)
81. L.P. Vergasova, S.K. Filatov, E. K. Serafimova et al., Proc. Miner. Soc. USSR **117**, 459 (1988) (in Russian)
82. F. Demartin, C.M. Gramaccioli, I. Campostrini et al., Canad. Miner. **46**, 693 (2008)
83. I.V. Pekov, M.E. Zelenski, N.V. Zubkova et al., Amer. Miner. **76**, 673 (2012)
84. L.P. Vergasova, S.K. Filatov, E. K. Serafimova et al., Dokl. Acad. Sci. USSR **275**, 714 (1984) (in Russian)
85. E.H. Nickel, P.J. Bridge, Miner. Mag. **42**, 37 (1977)
86. G.G. Urazov, N.I. Bashilova, Rus. J. Inorg. Chem. **2**, 1922 (1957) (in Russian)
87. L.P. Vergasova, S.K. Filatov, E.K. Serafimova et al., Dokl. Acad. Sci. USSR **299**, 961 (1988) (in Russian)
88. M.G. Gorskaya, L.P. Vergasova, S.K. Filatov et al., Proc. Rus. Miner. Soc. **1**, 95 (1995) (in Russian)
89. L.P. Vergasova, S.K. Filatov, M.G. Gorskaya et al., Proc. Miner. Soc. USSR **118**(1), 70 (1989) (in Russian)

90. E. Staritzky, A.L. Truitt, in *The actinide elements*, ed. by G.T. Seaborg, J.J. Katz (McGraw-Hill, New York, 1954)
91. A.R. Kampf, P.J. Dunn, E.E. Foord, Amer. Miner. **74**, 927 (1989)
92. C.S. Hurlbut, L.F. Aristarain, Amer. Miner. **54**, 1519 (1969)
93. A. Livingstone, H. Sarp, Miner. Mag. **48**, 277 (1984)
94. S.P. Altaner, J.J. Fitzpatrick, M.D. Krohn et al., Amer. Miner. **73**, 145 (1988)
95. S.V. Krivovichev, L.P. Vergasova, S.N. Britvin et al., Canad. Miner. **45**, 921 (2007)
96. E.P. Shcherbakova, L.F. Bazhenova, Proc. Miner. Soc. USSR **118**, 84 (1989) (in Russian)
97. N.M. Selivanova, A.I. Mayer, K.K. Samplavskaya, Rus. J. Inorg. Chem. **7**, 1074 (1962) (in Russian)
98. N.M. Selivanova, V.A. Shnaider, I.S. Strel'tsov, Rus. J. Inorg. Chem. **5**, 2272 (1960) (in Russian)
99. N.M. Selivanova, V.A. Shnaider, I.S. Strel'tsov, Rus. J. Inorg. Chem. **5**, 2269 (1960) (in Russian)
100. J. Goni, C. Guillemin, Bull. Soc. France Min. Crist. **76**, 422 (1953)

Chapter 9
Refractive Indices in the Coordination Compounds of Group 11–14 Metals

Besides RIs, these tables contain the densities (measured by pycnometry) for new compounds synthesised or characterised by the authors. Abbreviations: A = NH$_3$, An = aniline, Az = anisidine, En = ethylenediamine, Py = pyridine, To = toluidine (Tables 9.1, 9.2 and 9.3).

Table 9.1 Refractive indices and density (ρ, g/cm^3) in compounds of Group 11 metals

Compounds	ρ	n_g	n_m	n_p	Compounds	ρ	n_g	n_m	n_p
KCu(CN)$_2^a$		1.718	1.705	1.589	KAu(CN)$_2$ [h]		1.6943	1.6005	
K$_3$Cu(CN)$_4^b$			1.552	1.544	KAuCl$_4$	3.79	1.924	1.830	1.700
Cs$_2$CuCl$_4^c$		1.678	1.648	1.625	KAuCl$_3$Br	4.05	1.998	1.934	1.755
CuPy$_2$Cl$_2^d$		1.75	?	1.60	KAuClBr$_3$	4.57	2.15	2.10	1.870
CuEn$_3$S$_2$O$_3$	1.62		1.628	1.605	KAuBr$_4$	4.84	2.24	2.18	1.943
CuEn$_2$S$_2$O$_3$	1.81	1.666	1.654	1.637	RbAuCl$_4$	4.02	1.902	1.758	1.752
CuEn$_2$SeSO$_3$	2.05	1.714	1.682	1.677	RbAuCl$_3$Br	4.25	1.957	1.885	1.855
CuW$_4$SO$_4^e$		1.547	1.515	1.513	RbAuCl$_2$Br$_2$	4.51	2.021	1.973	1.944
KAg(CN)$_2^a$		1.6035	1.4915		RbAuClBr$_3$	4.70	2.12	2.006	1.960
K$_2$Ag(SCN)$_4^f$			1.67–1.68		RbAuBr$_4$	4.92	2.16	2.020	1.968
HAu(CN)$_2^g$		1.98	1.96	1.95	AuCl$_3$Py	2.87	1.945	1.750	1.646
LiAu(CN)$_2^h$			2.33	1.975	AuBr$_3$Py	3.44	>2.07	?	1.689
NaAu(CN)$_2^h$		1.833	1.823						

[a][1], [b][2], [c][3], [d][4], [e][5], [f][6], [g][7] and [h][8]

© The Author(s) 2016
S.S. Batsanov et al., *Refractive Indices of Solids*,
SpringerBriefs in Applied Sciences and Technology,
DOI 10.1007/978-981-10-0797-2_9

Table 9.2 Refractive indices and density (ρ, g/cm³) in compounds of Group 12 metals

Compounds	ρ	n_g	n_m	n_p	Compounds	ρ	n_g	n_m	n_p
ZnA_2Cl_2	2.091	1.630	1.603	1.585	$(NH_4)_2ZnCl_4$			1.5055	
ZnA_2Br_2	2.841	1.7072	1.6723	1.6455	Cs_2ZnCl_4	3.329	1.600	1.590	1.589
$ZnPy_2Cl_2$	1.564	1.643	1.631	1.552	Rb_3ZnBr_5	3.708		1.656	
$ZnPy_2Br_2$	1.909	1.662	1.646	1.595	$(NH_4)_2[CdA_2(H_2O)_2](SO_4)_2$[a]		1.514	1.491	
$ZnAn_2Cl_2$	1.56	1.730	1.628	1.597	$(NH_4)_2[CdA_3H_2O](SO_4)_2$[a]		1.499	1.493	1.483
$ZnAn_2Br_2$	1.880	1.754	1.652	1.628	CdA_4SeSO_3	2.37	1.683	1.640	1.611
$ZnAn_2I_2$	2.196	1.769	1.706	1.689	$CdEn_3S_2O_3$	1.80		1.623	1.598
$Zn(o\text{-}To)_2Cl_2$	1.470		1.662	1.576	$CdEn_3SeSO_3$	1.90		1.634	1.611
$Zn(o\text{-}To)_2Br_2$	1.774		1.692	1.601	K_4CdCl_6			1.5907	1.5906
$Zn(o\text{-}To)_2I_2$	2.09		1.755	1.660	Rb_4CdCl_6			1.580	
$Zn(p\text{-}To)_2Cl_2$	1.462	1.686	1.618	1.589	$(NH_4)_4CdCl_6$			1.6042	1.6038
$Zn(p\text{-}To)_2Br_2$	1.816	1.709	1.687	1.632	Cs_4CdCl_6		1.748	?	1.740
$Zn(o\text{-}Az)_2Cl_2$	1.543	1.706	1.656	1.551	$K_2Cd(NO_2)_4$		1.608	1.565	1.556
$Zn(o\text{-}Az)_2Br_2$	1.823	1.737	1.693	1.576	$K_2Cd(CN)_4$	1.824		1.4213	
$Zn(p\text{-}Az)_2Cl_2$	1.542	1.680	1.617	1.592	$K_2Hg(CN)_4$			1.458	
$Zn(o\text{-}Az)_2Br_2$	1.84	1.703	1.654	1.637	$KHg(SCN)_3$		1.880	1.843	1.730
$ZnEn_3S_2O_3$	1.60		1.614	1.597	$K_2Hg(SCN)_4$		1.90	1.80	1.645
$ZnEn_3SeSO_3$	1.77		1.627	1.605	$CsHgCl_3$			1.791	
$K_2Zn(CN)_4$	1.673		1.413		Cs_2HgI_4[b]	4.358	2.72	2.69	2.67

[a][9] and [b][10]

Table 9.3 Refractive indices of the boron, gallium, indium and tin complex compounds

Compounds	n_g	n_m	n_p	Compounds	n_g	n_m	n_p
NaBH$_4^a$		1.547		(NHMe$_3$)$_3$InCl$_6^b$	1.549	1.546	
KBH$_4^a$		1.490		(NH$_2$Me$_2$)$_4$InCl$_7^b$	1.574	1.560	1.550
RbBH$_4^a$		1.487		(NH$_3$Me)$_4$InCl$_7^b$	1.595	1.582	
CsBH$_4^a$		1.498		(NMe$_4$)$_2$InBr$_5^b$		1.599	
NaBF$_4$	1.307	1.301	1.301	Rb$_2$[In(OH)$_5$(H$_2$O)]	1.573	?	1.546
KBF$_4$	1.3247	1.3245	1.3239	(N$_2$H$_5$)$_2$SnF$_6^d$	1.453	1.451	1.442
CsBF$_4$		1.36		Rb$_2$SnCl$_6^e$		1.658	
Mn(BF$_4$)$_2$	1.359	?	1.346	Cs$_2$SnCl$_6^e$		1.647	
KGa(OH)$_4^c$	1.509	?	1.485	(Me$_4$N)$_2$SnCl$_6^f$		1.511	
Sr$_3$[Ga(OH)$_6$]$_2$		1.625		(Et$_4$N)$_2$SnCl$_6^f$	1.565	1.563	1.554
(NMe$_4$)$_2$InCl$_5^b$	1.555	1.550	1.54+	(C$_6$H$_{13}$N$_2$)$_2$SnBr$_6^h$	1.758	1.654	1.623
(NEt$_4$)$_2$InCl$_5^b$	1.583	1.565		Na$_2$Sn(OH)$_6^g$	1.582	1.568	

^{a}[11], ^{b}[12], ^{c}[13], ^{d}[14], ^{e}[15], ^{f}[16], ^{g}[17], ^{h}[18]

References

1. E. Staritzky, D.I. Walker, Anal. Chem. **28**, 419 (1956)
2. E.G. Cox, W. Wardlaw, K.C. Webster, J. Chem. Soc. 775 (1936)
3. D.P. Mellor, F.M. Quodling, Z. Krist. **95**, 315 (1936)
4. E.G. Cox, E. Sharratt, W. Wardlaw, K.C. Webster, J. Chem. Soc. 129 (1936)
5. I.G. Druzhinin, O.A. Kosyakina, Rus. J. Inorg. Chem. (in Russian) **6**, 1702 (1961)
6. V.J. Occleshaw, J. Chem. Soc. 2404 (1932)
7. R.A. Penneman, E. Staritzky, L.H. Jones, J. Am. Chem. Soc. **78**, 62 (1956)
8. R.A. Penneman, E. Staritzky, J. Inorg. Nucl. Chem. **7**, 45 (1958)
9. G.G. Urazov, A.K. Kirakosyan, Proc. Inst. Gen. Inorg. Chem. USSR (in Russian) **22**, 261 (1953)
10. A.A. Lavrentyev, B.V. Gabrelian, V.T. Vu et al., Optics Mater. **42**, 351 (2015)
11. M.D. Banus, R.W. Bragdon, A.A. Hinckley, J. Am. Chem. Soc. **76**, 3848 (1954)
12. J.B. Ekeley, H.A. Potratz, J. Am. Chem. Soc. **58**, 907 (1936)
13. B.I. Ivanov-Emin, Ya.I. Rabovik, Rus. J. Gen. Chem. (in Russian) **17**, 1061 (1947)
14. W. Pugh, J. Chem. Soc. 1934 (1953)
15. O.M. Ansheles, T.N. Burakova, *Crystallooptic as Foundation of Microchemical Analysis, L.* (Leningrad University Press, 1948)
16. E. Staritzky, J. Singer, Acta Cryst. **5**, 536 (1952)
17. J. Krc, Analyt. Chem. **23**, 675 (1951)
18. W. Pugh, J. Chem. Soc. 2491 (1953)

Chapter 10
Refractive Indices of Coordination Compounds of d- and f-Metals

Besides RIs, the tables below give the densities for compounds first synthesised and characterised by the authors. Where appropriate, the coordination polyhedra are described in terms of Werner coordinates; e.g. a $M(XY)X_2Y_2$ formula implies a coordination octahedron with the axes (coordinates) X–M–Y, X–M–X and Y–M–Y. Abbreviations: A = NH_3, Py = pyridine, En = $H_2NCH_2CH_2NH_2$ (Tables 10.1, 10.2, 10.3, 10.4, 10.5, 10.6 and 10.7).

Table 10.1 Refractive indices of lanthanide and actinide compounds

Compounds	n_g	n_m	n_p	Compounds	n_g	n_m	n_p
$K_2Ce(NO_3)_6^a$	1.717	1.681	1.659	$K_2Pu(NO_3)_6^a$	1.658	1.633	1.625
$Rb_2Ce(NO_3)_6^a$	1.708	1.665	1.654	$(NH_4)_2Pu(NO_3)_6^a$	1.663	1.639	1.633
$(NH_4)_2Ce(NO_3)_6^a$	1.705	1.673	1.659	$Rb_2Pu(NO_3)_6^a$	1.660	1.630	1.621
$Cs_2Ce(NO_3)_6^a$	1.704	1.663	1.659	$Cs_2Pu(NO_3)_6^a$	1.657	1.621	1.620
$Tl_2Ce(NO_3)_6^a$	1.78$_4$	1.76$_7$	1.75$_3$	$Tl_2Pu(NO_3)_6^a$	1.735	1.721	1.716
$Na_4Ce(SO_4)_4^b$	1.666	?	1.621	$(NMe_4)_2UCl_6^c$		1.511	
$(NH_4)_4Ce(SO_4)_4^b$	1.603	1.581	1.570	$(NMe_4)_2PuCl_6^c$		1.526	
$Rb_2Th(NO_3)_6^a$	1.609	1.586	1.576	$(NEt_4)_2UCl_6^c$	1.556	1.555	1.548
$(NH_4)_2Th(NO_3)_6^a$	1.613	1.599	1.588	$(NEt_4)_2PuCl_6^b$	1.569	1.568	1.560
$Cs_2Th(NO_3)_6^a$	1.613	1.586	1.579				

(a) [1], (b) [2], (c) [3]

© The Author(s) 2016
S.S. Batsanov et al., *Refractive Indices of Solids*,
SpringerBriefs in Applied Sciences and Technology,
DOI 10.1007/978-981-10-0797-2_10

Table 10.2 Refractive indices of iron complexes

Compounds	n_g	n_m	n_p	Compounds	n_g	n_m	n_p
$Fe(C_5H_5)_2$	1.737	1.733	1.648	$KInFe(CN)_6^a$		1.600	
$(NH_4)_2FeCl_4$		1.644		$RbInFe(CN)_6^a$		1.584	
$(NH_4)_3FeF_6$		1.442		$CsInFe(CN)_6^a$		1.605	
$H_3Fe(CN)_6$	1.755	1.678		$Zn_3[Fe(CN)_6]_2$		1.549	
$K_3Fe(CN)_6$	1.583	1.569	1.566	$Cd_3[Fe(CN)_6]_2$		1.539	
$(NH_4)_3Fe(CN)_6$		1.581		$In_4[Fe(CN)_6]_3^a$		1.576	
$H_4Fe(CN)_6$		1.644		$[FeEn_3]SeSO_3$		1.629	1.606
$K_4Fe(CN)_6$	1.591	1.589	1.585	$[FeEn_3]S_2O_3$		1.616	1.598

^{a}[4]

Table 10.3 Refractive indices of cobalt complexes

Compounds	n_g	n_m	n_p	Compounds	n_g	n_m	n_p
$[CoA_6][Co(CO_3)_3]^a$	1.737	?	1.603	$K[Co(NO_2)_4A_2]^b$	1.752	1.716	1.708
$CoPy_3(NO_2)_3^c$	1.754	1.720	1.704	$Cs_2CoCl_4^d$	1.584	1.579	1.575
$[CoA_6][TlCl_6]^e$		~1.76		α-$CoPy_2Cl_2^f$	1.767	1.750	1.580
$[CoA_6][TlBr_6]^e$		~1.80		β-$CoPy_2Cl_2^f$	1.652	1.620	1.551
$CoPy_4SCN_2^g$	1.718	1.672	1.628	$CoPy_2Br_2^f$	1.67	1.65	1.57
$[CoA_6]Cl_3$		1.701		$NaK_2Co(NO_2)_6^h$		1.730	
$[CoA_6]NO_3$	1.610	1.602	1.599	$Na_3Co(NO_2)_6$		1.787	
$[CoA_6][Co(NO_2)_6]$		1.796		$K_3Co(NO_2)_6$		1.723	
$[CoA_6][CoA_2(NO_2)_4]_3$	1.797	1.748	1.735	$(NH_4)_3Co(NO_2)_6$		1.742	
$[CoA_4(ANO_2)]Cl$	1.740	1.700	1.630	$K_2[Co(NCO)_4]$		1.519	1.479
$[CoA_4(ANO_2)]NO_3$	1.676	1.561		$K[CoA_2(NO_2)_4]$	1.755	1.715	1.700
$[CoA_2(ANO_2)_2]NO_3$	1.742	1.680	1.643	$NH_4[CoA_2(NO_2)_4]$	1.770	1.715	1.700
$[CoA_4(NO_2)_2]NO_3$	1.840	1.652	1.620	$CoA_2(ANO_2)(NO_2)_2$	1.794	1.749	1.734
$[CoA_2(ANO_2)_2]Cl$	1.749	1.692		$CoA_2(ANO_2)(NO_2)_2$	1.756	1.736	1.726
$[CoA_4(NO_2)_2]Cl$	1.838	1.720	1.580	$[CoA_4(NO_2)_2]_3[Co(NO_2)_6]$	1.765	1.753	1.708
$[CoA_4(ACl)]Cl_2$	1.711	1.699	1.686	$[CoA_5NO_2]_3[Co(NO_2)_6]$		1.763	
$[CoA_4(ACl)]NO_3)_2$	1.655	1.540		$[CoA_5NO_2]_3[CoA_2(NO_2)_4]_2$	1.784	1.732	1.720
$[Co(H_2O)_6](NO_3)_2^i$	1.547	1.52	1.38	$[CoEn_3]S_2O_3$		1.616	1.599
				$[CoEn_3]SeSO_3$		1.633	1.610

(a) [5], (b) [6], (c) [7], (d) [8], (e) [9], (f) [10], (g) [11], (h) [12], (i) [13]

Table 10.4 Refractive indices of nickel complexes

Compounds	n_g	n_m	n_p	Compounds	n_g	n_m	n_p
NiPy$_2$Cl$_2^a$	1.710	1.660	1.646	NiEn$_2$(SCN)$_2$	1.718	1.706	1.591
NiA$_3$(SCN)$_2^b$	1.748	1.732	1.568	K$_2$Ni(SCN)$_4$		1.54	
NiA$_4$(NO$_2$)$_2^b$	1.660	1.598	1.491	NH$_4$[NiA$_3$(SCN)$_3$]^b		1.658	1.650
NiA$_4$(SCN)$_2^b$	1.674	1.618	1.561	[NiEn$_3$]S$_2$O$_3$		1.624	1.604
NiPy$_4$Br$_2^b$	1.770	1.692	1.682	[NiEn$_3$]SeSO$_3$		1.635	1.612
NiPy$_4$(SCN)$_2$ ^{b}	1.718	1.680	1.642				

(a) [14], (b) [15]

Table 10.5 Refractive indices of the complexes of ruthenium, rhodium and palladium

Compounds	n_g	n_m	n_p	Compounds	n_g	n_m	n_p
K$_2$[Ru(NO)Cl$_5$]^a	≥ 1.78	?	1.745	(NH$_4$)$_2$PdCl$_4^b$		1.736	1.544
K$_2$[Ru(NO)(OH)(NO$_2$)$_4$]^c	1.737	1.650	1.609	Cs$_2$PdCl$_4^d$		1.716	1.556
(NH$_4$)$_2$[Ru(NO)(OH)Cl$_4$]^e	1.798	1.779	1.757	Tl$_2$PdCl$_4^d$		~2.21	1.97
[RuA$_4$(NO)(OH)]Cl$_2^f$	1.830	1.708	1.661	Cs$_2$PdBr$_4^d$		~1.93	
Na(NH$_4$)$_2$[Rh(NO$_2$)$_6$]^g		1.680		K$_2$Pd(NO$_2$)$_4^d$	1.694	?	1.598
K$_3$Rh(NO$_2$)$_6^h$		1.686		K[Pd(C$_7$H$_6$O$_2$N)(C$_7$H$_5$O$_2$N)]^i	>1.78	?	1.542
RhA$_3$(NO$_2$)$_3^h$	1.780	1.720	1.700	NH$_4$[Pd(C$_7$H$_6$O$_2$N)(C$_7$H$_5$O$_2$N)]^i	>1.78	?	1.532
RhA$_3$Cl$_3^h$	>1.78	>1.78	1.766	Pd(C$_5$H$_5$O$_2$N)$_2$Cl$_2$ ^{j}	>1.78	?	1.53
RhA$_3$(NO$_2$)$_3^h$	1.736	1.732	1.722	PdA$_2$Cl$_2$	1.788	1.785	1.645
Rh(C$_7$H$_6$O$_2$N)$_3^i$		1.75		PdA$_2$I$_2^d$	>1.78	?	1.777
[RhA$_6$]Cl$_3^h$		1.679		[PdA$_4$][Pd(NO$_2$)$_3$Cl]^k	1.710	?	1.612
K$_2$PdCl$_4^d$		1.714	1.522	[PdA$_2$(C$_5$H$_4$O$_2$N)$_2$]^j		1.76	
Rb$_2$PdCl$_4^d$		1.712	1.525	[PdA$_2$(C$_5$H$_5$O$_2$N)$_2$]C$_2$O$_4^j$	>1.78	?	1.55

(a) [16], (b) [17], (c) [18], (d) [19], (e) [20], (f) [21], (g) [22], (h) [23], (i) [24], (j) [25], (k) [26]

Table 10.6 Refractive indices of platinum(II) complexes

Compounds	n_g	n_m	n_p	Compounds	n_g	n_m	n_p
$K_2PtCl_4^a$		1.682	1.552	$PtA_2(NO_2Cl)^a$	1.790	1.786	1.764
$(NH_4)_2PtCl_4^b$		1.706	1.574	$PtA_2(NO_2Br)^c$	1.849	1.822	1.778
$K_2PtCl_2Br_2$		1.726	1.550	$PtA_2(NO_2)_2^d$	1.815	1.779	1.531
$K_2PtBr_4^d$		1.782	1.574	$Pt(ANO_2)_2^d$	1.791	1.742	1.712
$K[PtACl(NO_2)_2]^g$	1.752	1.638	1.630	$Pt(ASCN)_2^e$	1.828	1.780	1.730
$K_2Pt(NO_2)_4^d$	1.668	1.643	1.574	$PtA_2(SCN)_2^f$	1.990	1.737	1.625
$NH_4PtACl_3^b$	1.743	1.662	1.576	$PtA_2C_2O_4^h$	1.83	1.670	1.612
$NH_4Pt(C_2H_4)Cl_3^c$	1.78	1.724	1.595	$[PtA_3Cl]Cl^d$	1.767	1.721	1.654
$Pt(ACl)(PyCl)^i$	>1.79	1.732	1.624	$[PtA_3Br]Br^d$	1.829	1.762	1.685
$PtAPyCl_2^i$	1.82	1.732	1.653	$[PtA_3NO_2]NO_2^d$	1.731	1.692	1.636
$Pt(APy)(NO_2)_2^a$	1.750	1.696	1.624	$[PtA_4][Pt(NO_2)_4]^c$	1.78	1.71	
$Pt(ACl)(COCl)^a$	1.790	1.745	1.722	$Pt(APy)(NO_2Cl)^c$	1.785	1.740	1.595
$Pt(ACl)(C_2H_4Cl)^a$	>1.785	1.722		$PtAPy(NO_2)_2^c$	1.750	1.696	1.624
$Pt(ABr)(C_2H_4Br)^a$	>1.790	1.770		$[PtA_4][PtCl_4]^c$	>1.853	1.770	
$Pt(ACl)(C_8H_8Cl)^a$	>1.79	>1.79	1.668	$[PtA_4][Pt(NO_2)_2Br_2]^j$	1.776	?	1.726
$Pt(ACl)_2^c$	1.812	1.790	1.745	$[PtA_4][Pt(NO_2Br)_2]^j$	1.813	?	1.737
$PtA_2Cl_2^c$	1.850	1.778	1.706	$PtPy_2Cl_2^c$	>1.78	1.770	1.574
$Pt(ABr)_2^k$	1.957	1.866	1.832	$Pt(PyCl)_2^c$	1.780	>1.70	1.620
$PtA_2Br_2^{dk}$	1.923	1.813	1.730	$Pt(PyCl)(C_2H_4Cl)^a$	1.97	1.704	1.682
$Pt(AI)_2^k$	1.951	1.877	1.800	$Pt(PyCl)(C_4H_6Cl)^a$	1.80	1.756	1.700
$Pt(ACl)(C_2H_4Br)^m$	>1.783	1.758		$Pt(N_2H_4Cl)_2^c$	>1.78	1.76	1.745
$Pt(AC_2H_4)(BrCl)^m$	>1.783	1.708		$Pt(ACl)(COCl)^l$	1.790	1.745	1.722
$Pt(ABr)(PyCl)^n$	>1.776	1.762	1.638	$Pt(NH_2OH)_2Cl_2$	>1.782	?	1.78
$Pt(ACl)(PyBr)^n$	>1.776	1.730	1.635	$Pt(NH_2OHCl)_2$	>1.782	?	1.76
$Pt(APy)(BrCl)^n$	>1.776	1.776	1.567	$K[Pt(NO_2)_2ACl]^p$	1.752	1.638	1.630
$[PtA_2PyNH_2OH][PtCl_4]^o$	1.776	1.752	1.698				

(a) [27], (b) [28], (c) [29], (d) [30], (e) [31], (f) [32], (g) [33], (h) [34], (i) [35], (j) [36], (k) [37], (l) [38], (m) [39], (n) [40], (o) [41], (p) [42]

Table 10.7 Refractive indices of platinum (IV) complexes $(Ma = CH_3NH_2)$

Compounds	n_g	n_m	n_p	Compounds	n_g	n_m	n_p
Li_2PtF_6		1.610	1.566	$Pt(ACl)_2(NO_2)_2$[g]	>1.78	?	1.682
Na_2PtF_6		1.469	1.448	$PtA_2(NO_2)_2Cl_2$	1.980	1.822	1.725
K_2PtF_6		1.532	1.498	$Pt(ANO_2)_2Cl_2$		1.784	
Rb_2PtF_6		1.541	1.508	$PtA_2(NO_2)_2BrCl$	1.950	1.835	1.752
Cs_2PtF_6		1.573	1.542	$Pt(ANO_2)_2BrCl$	1.892	1.827	1.782
K_2PtCl_6		1.825		$PtA_2(NO_2)_2Br_2$	1.942	1.866	1.790
$(NH_4)_2PtCl_6$		1.843		$Pt(ANO_2)_2Br_2$	1.968	1.857	1.822
$(NEt_4)_2PtCl_6$[a]		1.620	1.613	$Pt(ANO_2)_2(OH)_2$	1.815	1.785	1.698
$[NH(CH_3)_3]PtCl_6$		1.600		$Pt(ANO_2)_2ClOH$	1.790	1.778	1.756
$Li_2Pt(CN)_6$		1.536	1.502	$Pt(ANO_2)_2BrOH$	1.881	1.823	1.756
$Na_2Pt(CN)_6$		1.568	1.549	$PtA_2(NO_2)_2NO_2Cl$[h]	1.779	1.762	1.745
$K_2Pt(CN)_6$		1.505	1.491	$Pt(ANO_2)_2NO_2Cl$	1.89	1.797	1.755
$Rb_2Pt(CN)_6$		1.489	1.478	$Pt(ANO_2)_2BrOH$	1.881	1.823	1.756
$Cs_2Pt(CN)_6$		1.597	1.560	$PtA_2(NO_2)_2Cl_2$	1.980	1.822	1.725
$K_2Pt(NO_2)_6$[b]		1.714		$Pt(ANO_2)_2Cl_2$		1.784	
$K_2Pt(NO_2)_4Cl_2$	1.780	1.730	1.716	$Pt(NO_2)_2(MaNO_2)_2$[i]	1.780	?	1.740
$K_2Pt(NO_2)_2(NO_2Cl)_2$	1.810	1.756	1.724	$Pt(NO_2NO_3)(MaNO_2)$[i]	1.754	1.734	1.694
$K_2Pt(NO_2Cl)_3$	1.880	1.791	1.740	$Pt(NO_2OH)(MaNO_2)_2$[i]	1.724	?	1.714
$K_2Pt(NO_2)_2(NO_2Cl)Cl_2$	1.791	1.761	1.720	$Pt(NO_2Cl)(MaNO_2)_2$[i]	1.767	?	1.740
$K_2Pt(NO_2Cl)_2Cl_2$	1.835	1.770	1.708	$[Pt(NO_2Cl)EnABr]Cl$[j]	1.78	?	1.662
$K_2Pt(NO_2)_2Cl_4$	1.796	1.761	1.720	$[Pt(NO_2Br)EnACl]Cl$[j]	>1.776	?	1.716
$Cs_2Pt(NO_2)_2(NO_2Cl)_2$	1.796	1.760	1.709	$[Pt(ACl)EnNO_2Br]NO_3$[k]	1.737	?	1.690
$Cs_2Pt(NO_2)_4Cl_2$	1.807	1.767	1.702	$[Pt(ABr)EnNO_2Cl]NO_3$[k]	1.776	?	1.675
$K_2(NO_2Br)_2Br_2$[c]	1.960	?	1.799	$Pt(APy)(NO_2Cl)Br_2$[l]	1.868	?	1.835
$K_2Pt(NO_2)_2Br_4$[c]	1.968	?	1.809	$Pt(APy)(NO_2Cl)BrI$[l]		1.910	
$K_2Pt(NO_2Br)_3$[c]	1.916	?	1.816	$Pt(ACl)(PyNO_2)BrI$[l]	1.920	?	1.788
$K_2Pt(NO_2)_2(NO_2Br)_2$[c]	1.906	?	1.811	$Pt(ACl)(PyNO_2)Br_2$[l]	1.920	?	1.780
$K_2Pt(NO_2)_2Br_2Cl_2$	1.908	1.801	1.778	$PtCl_2(EnNO_2Br)$[j]	1.780	?	1.767
$K_2Pt(SCN)_6$		1.890	1.820	$[Pt(BrCl)(EnANO_2)]Cl$[j]	>1.776	?	1.73
$[PtA_6]Cl_4$[d]		1.727	1.724	$[PtBr_2(EnANO_2)]Cl$[j]	>1.78	1.776	1.743
$[PtA_4Cl_2]Cl_2$	1.996	1.745		$[Pt(MaBr)EnNO_2Cl]NO_3$[m]	1.74	?	1.690
$[PtA_4(ACl)]SO_4Cl$[e]	1.706	1.674		$[Pt(MaCl)EnNO_2Br]NO_3$[m]	1.728	?	1.698
PtA_2Cl_4	2.002	1.856		$[Pt(MaCl)EnNO_2Br]Cl$[m]	1.740	?	1.714
$Pt(ACl)_2Cl_2$	1.994	1.870	1.842	$[Pt(MaBr)EnNO_2Cl]Cl$[m]	1.718	?	1.71
$PtA_2Cl_2(OOH)_2$[f]	1.756	1.730	1.690				

[a][3], [b][43], [c][36], [d][29], [e][44], [f][17], [g][42], [h][45], [i][46], [j) [47], [k][48], [l][49], [m][50]

References

1. E. Staritzky, A.L. Truitt, in *The actinide elements*, ed. by G.T. Seaborg, J.J. Katz (McGraw-Hill, New York 1954)
2. V.A. Golovnya, L.A. Pospelova, Rus. J. Inorg. Chem. (in Russian) **6**, 1574 (1961)
3. E. Staritzky, J. Singer, Acta Cryst. **5**, 536 (1952)
4. E.N. Deichman, *Proc. Acad. Sci. USSR, Chem.* (in Russian) **9**, 1013 (1957)
5. V.A. Golovnya, L.A. Kokh, Rus. J. Inorg. Chem. (in Russian) **6**, 1774 (1961)
6. G.B. Bokii, E.A. Gilinskaya, *Proc. Acad. Sci. USSR, Chem.* (in Russian) **2**, 238 (1953)
7. A.V. Babaeva, I.B. Baranovskiy, G.G. Afanas'eva, *Dokl.Acad. Sci. USSR* (in Russian) **143**, 587 (1962)
8. M.A. Porai-Koshits, Crystallography (in Russian) **1**, 291 (1956)
9. T. Watanabe, M. Atoji, C. Okazaki, Acta Cryst. **3**, 405 (1950)
10. M.N. Lyashenko, Crystallography (in Russian) **1**, 361 (1956)
11. A.S. Antsyshkina, Crystallography (in Russian) **3**, 742 (1958)
12. O.M. Ansheles, T.N. Burlakova, *Crystallooptics as foundation of microanalysis, L.* (Leningrad University Press, 1948)
13. A. Yayaraman, Proc. Indian Acad. Sci. **A47**, 147 (1958)
14. A.V. Babaeva, C. Shou-Gyan, Rus. J. Inorg. Chem. (in Russian) **5**, 2167 (1960)
15. M.A. Porai-Koshits, E.K. Yukhno, A.S. Antsyshkina, L.M. Dikareva, Crystallography (in Russian) **2**, 371 (1957)
16. T.S. Khodasheva, G.B. Bokii, J. Struct. Chem. **1**, 138 (1960)
17. G.B. Bokii, M.N. Lyashenko, *Proc. Inst. Crystallogr.* (in Russian) **3**, 37 (1947)
18. G.B. Bokii, Van Angpu, T.S. Khodasheva, *J. Struct. Chem.* **3**, 149 (1962)
19. T.N. Burlakova, *Bull. Leningrad Univ.* (in Russian) **178**, 157 (1954)
20. N.A. Parpiev, M.A. Porai-Koshits, Crystallography (in Russian) **4**, 30 (1959)
21. G.B. Bokii, N.A. Parpiev, Crystallography (in Russian) **2**, 691 (1957)
22. G.B. Bokii, L.A. Popova, *Proc. Acad. Sci. USSR, Chem.* (in Russian) **2**, 89 (1945)
23. M.N. Lyashenko, *Proc. Inst. Crystallogr.* (in Russian) **7**, 67 (1952)
24. N.K. Pshenitsin, G.A. Nekrasova, Ann. Inst. Platine (USSR) **30**, 159 (1955)
25. N.K. Pshenitsin, G.A. Nekrasova, Ann. Inst. Platine (USSR) **30**, 143 (1955)
26. I.I. Chernyaev, G.S. Muraveiskaya, Rus. J. Inorg. Chem. (in Russian) **2**, 772 (1957)
27. G.B. Bokii, M.N. Lyashenko, *Proc. Inst. Crystallogr.* (in Russian) **3**, 21 (1947)
28. G.B. Bokii, E.E.Burovaya, *Proc. Inst. Crystallogr.* (in Russian) **3**, 47 (1947)
29. M.M. Yakshin, Ann. Inst. Platine (USSR) **21**, 146 (1948)
30. M.N. Lyashenko, Proc. Inst. Crystallogr. (in Russian) **9**, 335 (1954)
31. Ya.Ya. Bleidelis, Crystallography (in Russian) **2**, 278 (1957)
32. Ya.Ya. Bleidelis, G.B. Bokii, *Crystallography* (in Russian), **2**, 281 (1957)
33. M.N. Lyashenko, Crystallography (in Russian) **1**, 361 (1956)
34. D.S. Flikkema, Acta Cryst. **6**, 37 (1953)
35. G.B. Bokii, E.E.Burovaya, *Proc. Inst. Crystallogr.* (in Russian) **2**, 111 (1940)
36. A.V. Babaeva, Van Yubin', *Rus. J. Inorg. Chem.* (in Russian) **6**, 1525 (1961)
37. M.M. Yakshin, V.M. Ezuchevskaya, Rus. J. Inorg. Chem. (in Russian) **2**, 555 (1957)
38. A.D. Gel'man, *Ann. Inst. Platine (USSR)* **18**, 50 (1945)
39. A.D. Gel'man, E.A. Meilakh, *Ann. Inst.Platine (USSR)* **20**, 21 (1947)
40. A.D. Gel'man, E.F. Karandasheva, L.N. Essen, *Ann. Inst.Platine (USSR)* **24**, 60 (1949)
41. V.I. Goremykin, K.A. Gladyshevskaya, Ann. Inst. Platine (USSR) **17**, 67 (1940)
42. I.I. Chernyaev, G.S. Muraveiskaya, Ann. Inst. Platine (USSR) **31**, 5 (1955)
43. I.I. Chernyaev, L.A. Nazarova, A.S. Mironova, Rus. J. Inorg. Chem. (in Russian) **6**, 2444 (1961)
44. A.M. Rubinshtein, Ann. Inst. Platine (USSR) **20**, 53 (1947)
45. E.E. Burovaya, *Proc.Inst.Crystallogr.* (in Russian) **5**, 197 (1949)
46. I.I. Chernyaev, G.S. Muraveiskaya, Rus. J. Inorg. Chem. (in Russian) **2**, 536 (1957)

47. I.I. Chernyaev, O.N. Adrianova, Ann. Inst. Platine (USSR) **23**, 9 (1949)
48. I.I. Chernyaev, O.N. Adrianova, Ann. Inst. Platine (USSR) **24**, 79 (1949)
49. L.N. Essen, A.D. Gel'man, *Dokl.Acad.Sci.USSR* (in Russian) **108**, 651 (1956)
50. I.I. Chernyaev, O.N. Adrianova, Ann. Inst. Platine (USSR) **31**, 26 (1955)

Part III
Crystallohydrates of Simple and Complex Compounds

Chapter 11
Crystallohydrates of Simple and Complex Compounds

Abbreviations: tr = triclinic and o = orthorhombic (Tables 11.1, 11.2, 11.3, 11.4, 11.5, 11.6, 11.7 and 11.8).

Table 11.1 Refractive indices of the hydrates in complex halides

Compounds	n_g	n_m	n_p	Compounds	n_g	n_m	n_p
LiI·3H$_2$O		1.655	1.625	ZnF$_2$·4H$_2$O[a]	1.47	?	1.46
NaBr·2H$_2$O	1.5252	1.5192	1.5128	MnCl$_2$·2H$_2$O	1.666	1.611	1.584
KF·2H$_2$O	1.363	1.352	1.345	MnCl$_2$·4H$_2$O	1.607	1.575	1.555
CuCl$_2$·2H$_2$O	1.742	1.684	1.644	FeCl$_2$·2H$_2$O[b]	1.703	1.633	1.605
Cu(OH)$_2$, H$_2$O		1.708	1.702	CoCl$_2$·2H$_2$O	1.721	1.662	1.626
BeCl$_2$·4H$_2$O[c]		1.513		NiCl$_2$·2H$_2$O	1.783	1.723	1.620
MgCl$_2$·6H$_2$O	1.528	1.507	1.495	NiCl$_2$·6H$_2$O	1.61	?	1.535
tr-CaCl$_2$·4H$_2$O	1.571	1.560	1.532	AlF$_3$·H$_2$O	1.511	1.490	1.473
o-CaCl$_2$·4H$_2$O	1.491	1.477	1.447	AlCl$_3$·6H$_2$O		1.560	1.507
CaCl$_2$·6H$_2$O		1.417	1.393	AlBr$_3$·6H$_2$O		1.605	1.555
SrCl$_2$·2H$_2$O	1.6172	1.5948	1.5942	EuCl$_3$·6H$_2$O[d]	1.5818	1.5788	1.5700
SrCl$_2$·6H$_2$O		1.5356	1.4857	GdCl$_3$·6H$_2$O	1.575	1.570	1.565
SrBr$_2$·6H$_2$O		1.557	1.535	SmCl$_3$·6H$_2$O	1.573	1.569	1.564
BaCl$_2$·2H$_2$O	1.660	1.646	1.641	PuCl$_3$·6H$_2$O[e]	1.597	1.596	1.582
BaBr$_2$·2H$_2$O	1.7441	1.7266	1.7129	InF$_3$·3H$_2$O[f]		1.450	1.425
KMgCl$_3$·6H$_2$O	1.4937	1.4753	1.4665	Li$_2$SiF$_6$·2H$_2$O	1.300	?	1.296
K$_2$CuCl$_4$·2H$_2$O		1.6485	1.6133	NaCaAlF$_6$·H$_2$O	1.420	1.413	1.411
K$_2$HgCl$_4$·H$_2$O	1.699	1.678	1.648	Cu(BF$_4$)$_2$·6H$_2$O		1.50	
(NH$_4$)$_2$MnCl$_4$·2H$_2$O		1.644	1.607	Sr$_3$Fe$_2$F$_{12}$·2H$_2$O	1.482	1.480	1.473
(NH$_4$)$_2$CoCl$_4$·2H$_2$O	1.682	?	1.640	BaFeF$_5$·H$_2$O	1.513	1.503	1.502
Cs$_2$MnCl$_4$·2H$_2$O	1.65	?	1.64	MgSiF$_6$·6H$_2$O	1.3602	1.3439	
(NH$_4$)$_2$CuCl$_4$·2H$_2$O		1.671	1.641	MnSiF$_6$·6H$_2$O	1.3742	1.3570	

(continued)

© The Author(s) 2016
S.S. Batsanov et al., *Refractive Indices of Solids*,
SpringerBriefs in Applied Sciences and Technology,
DOI 10.1007/978-981-10-0797-2_11

Table 11.1 (continued)

Compounds	n_g	n_m	n_p	Compounds	n_g	n_m	n_p
$NaAuCl_4 \cdot 2H_2O$	1.75	?	1.545	$FeSiF_6 \cdot 6H_2O$	1.3848	1.3638	
$KAuCl_4 \cdot 2H_2O$	1.69	1.56	1.55	$NiSiF_6 \cdot 6H_2O$	1.4066	1.3910	
$K_2FeCl_5 \cdot H_2O$	1.80	1.75	1.715	$CoSiF_6 \cdot 6H_2O$	1.3872	1.3817	
$(NH_4)_2FeCl_5 \cdot H_2O$	1.814	1.775	1.750	$CuSiF_6 \cdot 6H_2O$		1.4092	1.4080
$(NH_4)_3RhCl_6 \cdot H_2O$	1.756	1.750	1.740	$ZnSiF_6 \cdot 6H_2O$	1.3992	1.3824	
$(NH_4)_3IrCl_6 \cdot H_2O$	1.718	1.714	1.706	$MgSnCl_6 \cdot 6H_2O$	1.597	1.5885	
$HfOCl_2 \cdot 8H_2O$		1.557	1.543	$Ca_2OCl_2 \cdot 2H_2O$		1.638	11.634
$Ca_2OBr_2 \cdot 3H_2O$	1.645	1.623		$Ca_2OCl_2 \cdot H_2O$		1.628	1.623
$Ca_4O_3I_2 \cdot 15H_2O$		1.575		$Ca_4O_3Cl_2 \cdot 15H_2O$	1.543	1.536	1.481
$Ca_4O_3Br_2 \cdot 15H_2O$		1.555		$H_5NiF_7 \cdot 6H_2O$	1.408	1.392	
$BaCdCl_4 \cdot 4H_2O$	1.653	1.646	1.610	$H_5CoF_7 \cdot 6H_2O$	1.399	1.384	
$ZrOCl_2 \cdot 8H_2O$	1.563	1.552		$H_5CuF_7 \cdot 6H_2O$	1.444	1.440	1.395

[a][1], [b][2], [c][3], [d][4], [e][5] and [f][6]

Table 11.2 Refractive indices of the hydrates of cyanide complexes

Compounds	n_g	n_m	n_p	Compounds	n_g	n_m	n_p
$MgPt(CN)_4 \cdot 7H_2O$	1.91	1.561		$LiKPt(CN)_4 \cdot 3H_2O$	2.0405	1.6217	1.6183
$CaNi(CN)_4 \cdot 5H_2O$	1.638	1.617	1.5405	$LiRbPt(CN)_4 \cdot 3H_2O$	1.9310	1.6233	1.6204
$CaPd(CN)_4 \cdot 5H_2O$	1.639	1.602	1.539	$NaKPt(CN)_4 \cdot 3H_2O$	1.90	1.61	1.609
$CaPt(CN)_4 \cdot 5H_2O$	1.767	1.644	1.623	$Na_2Pt(CN)_4 \cdot 3H_2O$	1.611	1.608	1.541
$Ca_2Fe(CN)_6 \cdot 12H_2O$	1.5961	1.5818	1.5700	$Na_3Fe(CN)_6 \cdot 2H_2O$	1.560	1.549	1.531
$SrNi(CN)_4 \cdot 5H_2O$	1.6235	1.612	1.492	$K_4Fe(CN)_6 \cdot 3H_2O$	1.580	1.575	1.570
$SrPd(CN)_4 \cdot 5H_2O$	1.612	1.6025	1.495	$K_2Ni(CN)_4 \cdot 4H_2O$	1.5955	1.5915	1.4657
$SrPt(CN)_4 \cdot 5H_2O$	1.637	1.613	1.547	$K_4Ru(CN)_6 \cdot 3H_2O$		1.5837	
$BaNi(CN)_4 \cdot 4H_2O$	1.658	1.658	1.569	$K_4Os(CN)_6 \cdot 3H_2O$		1.6071	
$BaPd(CN)_4 \cdot 4H_2O$	1.651	1.646	1.583	$(NH_4)_4Fe$ $(CN)_6 \cdot 1.5H_2O$		1.590	
$BaPt(CN)_4 \cdot 4H_2O$	1.919	1.674	1.666	$Y_2Pt_3(CN)_{12} \cdot 21H_2O$	2.055	1.60	1.591
$La[Ag(CN)_2]_3 \cdot 3H_2O^a$	1.699	?	1.660	$Ce_2Pt_3(CN)_{12} \cdot 18H_2O$	1.68	1.66	1.65
$Pr[Ag(CN)_2]_3 \cdot 3H_2O^a$	1.717	?	1.688	$K_3[Ni(CN)_3S] \cdot H_2O^b$	1.603	?	1.490
$Nd[Ag(CN)_2]_3 \cdot 3H_2O^a$	1.718	?	1.689	$K_2CaFe(CN)_6 \cdot 3H_2O^c$		1.664	
$Gd[Ag(CN)_2]_3 \cdot 3H_2O^a$	1.719	?	1.689	$K_2BaFe(CN)_6 \cdot 3H_2O^c$		1.630	

[a][7], [b][8], [c][9]

Table 11.3 Refractive indices of the hydrates of nitrates and carbonates

Compounds	n_g	n_m	n_p	Compounds	n_g	n_m	n_p
$LiNO_3 \cdot H_2O$	1.523	1.490	1.365	$Mg_3Nd_2(NO_3)_{12} \cdot 24H_2O$		1.5266	1.5192
$Mg(NO_3)_2 \cdot 6H_2O$	1.506	1.506	1.34	$Sc(NO_3)_3 \cdot 4H_2O^a$	1.568	1.497	1.438
$Ca(NO_3)_2 \cdot 4H_2O$	1.504	1.498	1.465	$Y(NO_3)_3 \cdot 4H_2O$	1.570	1.528	1.420
$Co(NO_3)_2 \cdot 6H_2O^b$	1.547	1.52	1.38	$ScOH(NO_3)_2 \cdot 3H_2O^a$	1.601	1.564	1.464
$MgCe(NO_3)_6 \cdot 8H_2O^c$	1.589	1.587	1.573	$La(NO_3)_3 \cdot 4H_2O$	1.592	1.568	1.521
$ZnCe(NO_3)_6 \cdot 8H_2O^c$	1.603	1.601	1.583	$Ce(NO_3)_3 \cdot 4H_2O$	1.601	1.570	1.526
$CoCe(NO_3)_6 \cdot 8H_2O^c$	1.600	1.598	1.577	$Pr(NO_3)_3 \cdot 4H_2O$	1.603	1.574	1.531
$NiCe(NO_3)_6 \cdot 8H_2O^c$	1.603	1.602	1.579	$Nd(NO_3)_3 \cdot 4H_2O$	1.555	1.540	1.409
$MgTh(NO_3)_6 \cdot 8H_2O^c$	1.526	1.525	1.509	$Sm(NO_3)_3 \cdot 4H_2O$	1.562	1.546	1.413
$ZnTh(NO_3)_6 \cdot 8H_2O^c$	1.538	1.537	1.519	$La(NO_3)_3 \cdot 6H_2O$	1.592	1.584	1.449
$CoTh(NO_3)_6 \cdot 8H_2O^c$	1.540	1.540	1.521	$Ce(NO_3)_3 \cdot 6H_2O$	1.603	1.592	1.452
$NiTh(NO_3)_6 \cdot 8H_2O^c$	1.548	1.547	1.527	$Pr(NO_3)_3 \cdot 6H_2O$	1.558	1.498	1.476
$MgPu(NO_3)_6 \cdot 8H_2O^c$	1.554	1.553	1.538	$Nd(NO_3)_3 \cdot 6H_2O$	1.560	1.499	1.475
$ZnPu(NO_3)_6 \cdot 8H_2O^c$	1.570	1.568	1.550	$Sm(NO_3)_3 \cdot 6H_2O$	1.562	1.504	1.479
$CoPu(NO_3)_6 \cdot 8H_2O^c$	1.568	1.567	1.550	$Ce(NO_3)_4 \cdot 5H_2O^d$	1.691	1.590	1.586
$NiPu(NO_3)_6 \cdot 8H_2O^c$	1.576	1.576	1.550	$Th(NO_3)_4 \cdot 5H_2O^e$	1.628	1.528	1.518
$Mg_3La_2(NO_3)_{12} \cdot 24H_2O$	1.5220	1.5150		$Pu(NO_3)_4 \cdot 5H_2O^e$	1.667	1.556	1.554
$Mg_3Ce_2(NO_3)_{12} \cdot 24H_2O$	1.5249	1.5176		$ZrO(NO_3)_2 \cdot 2H_2O$	1.56	1.55	
$Mg_3Pr_2(NO_3)_{12} \cdot 24H_2O$	1.5255	1.5182		$Zr(NO_3)_4 \cdot 5H_2O$	1.61	1.60	
$HNa_3(CO_3)_2 \cdot 2H_2O$	1.540	1.492	1.412	$MgCO_3 \cdot 3H_2O$	1.527	1.503	1.417
$HKMg(CO_3)_2 \cdot 4H_2O$	1.542	1.51	1.430	$MgCO_3 \cdot 5H_2O$	1.508	1.469	1.456
$Na_2CO_3 \cdot H_2O$	1.524	1.506	1.420	$CaCO_3 \cdot 6H_2O$	1.545	1.535	1.460
$Na_2CO_3 \cdot 2.5H_2O$	1.547	1.492	1.435	$Mg_4(OH)_2(CO_3)_3 \cdot 3H_2O$	1.545	1.527	1.523
$Na_2CO_3 \cdot 10H_2O$	1.440	1.425	1.405	$Ag_2CO_3 \cdot 4NH_3 \cdot H_2O$	1.68	1.66	1.66

(continued)

Table 11.3 (continued)

Compounds	n_g	n_m	n_p
$Na_2CaCO_3 \cdot 2H_2O$	1.5751	1.5095	1.5043
$K_2MgCO_3 \cdot 4H_2O$	1.535	1.485	1.465
$Na_2CaCO_3 \cdot 5H_2O$	1.5233	1.5156	1.4435
$Na_2CuCO_3 \cdot 3H_2O$	1.576	1.530	1.483
$Na_6BaTh(CO_3)_6 \cdot 6H_2O$[k]	1.587	1.574	
$Na_3Y(CO_3)_3 \cdot 3H_2O$[l]	1.531	1.529	1.528
$NaY(CO_3)_2 \cdot 6H_2O$[n]	1.571	1.498	1.480
$NaCa_3(CO_3)_2F_3 \cdot H_2O$[o]	1.563	1.538	
$Na_3Y(CO_3)_3 \cdot 3H_2O$[q]		1.521	1.497
$NaGaOCO_3 \cdot H_2O$[t]	1.574	1.536	1.530
$KGaOCO_3 \cdot H_2O$[t]	1.582	1.570	1.565
$RbGaOCO_3 \cdot H_2O$[u]	1.601	?	1.569
$CsGaOCO_3 \cdot H_2O$[u]	1.618	?	1.578
$Na_6Th(CO_3)_5 \cdot 12H_2O$[x]	1.503	1.490	1.479

Compounds	n_g	n_m	n_p
$Ca_2SO_4CO_3 \cdot 4H_2O$[f]	1.531	1.518	1.516
$MgOH(CO_3)_2 \cdot 4H_2O$[g]	1.522	1.521	1.515
$Sr_5Zr_2(CO_3)_9 \cdot 4H_2O$[h]	1.654	1.646	1.553
$Ca_3(CO_3)_2Cl_2 \cdot 6H_2O$[i]	1.548	1.538	1.480
$CaY_2(CO_3)_4 \cdot 4H_2O$[j]	1.626	1.612	1.584
$SrLaOH(CO_3)_2 \cdot H_2O$[m]	1.731	1.717	1.640
$PbCr_2(OH)_4(CO_3)_2 \cdot H_2O$[p]	1.842	1.802	1.704
$Ba_2Al_4(CO_3)_4 \cdot 3H_2O$[r]	1.601	?	1.518
$KAlOCO_3 \cdot H_2O$[s]	1.544	1.540	1.502
$CsAlOCO_3 \cdot 2H_2O$[s]	1.547	?	1.536
$BaAl_2(OH)_4(CO_3)_2 \cdot 3H_2O$[v]	1.595	1.594	1.502
$Mg_4Al_2(OH)_{12}CO_3 \cdot 3H_2O$[w]		1.533	
$Mn_4Al_2(OH)_{12}CO_3 \cdot 3H_2O$[w]		1.587	1.547

a[10], b[11], c[5], d[12], e[13], f[14], g[15], h[16], i[17], j[18], k[19], l[20], m[21], n[22], o[23], p[24], q[25], r[26], s[27], t[28], u[29], v[30], w[31], x[32]

Table 11.4 Refractive indices in the hydrates of sulphates

Compounds	n_g	n_m	n_p	Compounds	n_g	n_m	n_p
$Li_2SO_4 \cdot H_2O$	1.488	1.477	1.459	$CoSO_4 \cdot 5H_2O$	1.550	1.548	1.530
$Na_2SO_4 \cdot 10H_2O$	1.398	1.396	1.394	$CoSO_4 \cdot 6H_2O$		1.495	1.460
$Cu(SO_4)_2 \cdot H_2O$	1.699	1.671	1.626	$CoSO_4 \cdot 7H_2O$	1.4885	1.4820	1.4728
$CuSO_4 \cdot 3H_2O$	1.618	1.577	1.554	$NiSO_4 \cdot 6H_2O$		1.5109	1.4873
$CuSO_4 \cdot 4H_2O$[a]	1.547	1.515	1.513	$NiSO_4 \cdot 7H_2O$	1.4923	1.4893	1.4693
$CuSO_4 \cdot 5H_2O$	1.5435	1.5368	1.5141	$Al_2(SO_4)_3 \cdot 15H_2O$	1.470	1.461	1.460
$CuSO_4 \cdot 7H_2O$	1.49	1.48	1.47	$Y_2(SO_4)_3 \cdot 8H_2O$	1.5755	1.549	1.5433
$Be(SO_4)_2 \cdot 4H_2O$		1.4702	1.4395	$La_2(SO_4)_3 \cdot 9H_2O$	1.569	1.564	
$Mg(SO_4)_2 \cdot H_2O$	1.596	1.525	1.523	$Ce(SO_4)_2 \cdot 4H_2O$[b]	1.784	1.689	1.6740 •
$MgSO_4 \cdot 4H_2O$	1.497	1.491	1.490	$Ce(SO_4)_2 \cdot 8H_2O$[b]	1.567	?	1.561
$MgSO_4 \cdot 5H_2O$	1.493	1.492	1.482	$Pr_2(SO_4)_3 \cdot 8H_2O$	1.5607	1.5494	1.5399
$MgSO_4 \cdot 6H_2O$	1.456	1.453	1.426	$Nd_2(SO_4)_3 \cdot 5H_2O$[c]	1.608	1.600	1.582
$Mg(SO_4)_2 \cdot 7H_2O$	1.4608	1.4554	1.4325	$Nd_2(SO_4)_3 \cdot 8H_2O$	1.5621	1.5505	1.5413
$CaSO_4 \cdot 2H_2O$	1.5296	1.5226	1.5205	$Sm_2(SO_4)_3 \cdot 8H_2O$	1.5629	1.5519	1.5427
$ZnSO_4 \cdot 6H_2O$		1.5291	1.5039	$Pu_2(SO_4)_3 \cdot 5H_2O$[c]	1.639	1.628	1.598
$ZnSO_4 \cdot 7H_2O$	1.4836	1.4801	1.4568	$In_2(SO_4)_3 \cdot 5H_2O$[d]	1.53	?	1.48
$MnSO_4 \cdot H_2O$	1.632	1.595	1.562	$Cr_2(SO_4)_3 \cdot 18H_2O$		1.564	
$MnSO_4 \cdot 4H_2O$	1.522	1.518	1.508	$Fe_2(SO_4)_3 \cdot 6H_2O$	1.657	1.635	1.605
$MnSO_4 \cdot 5H_2O$	1.514	1.508	1.495	$Fe_2(SO_4)_3 \cdot 7H_2O$	1.640	1.586	1.572
$FeSO_4 \cdot H_2O$	1.663	1.623	1.591	$Fe_2(SO_4)_3 \cdot 9H_2O$	1.572	1.536	
$FeSO_4 \cdot 4H_2O$	1.537	1.535	1.533	$Fe_3(SO_4)_4 \cdot 14H_2O$[e]	1.583	1.571	1.524
$FeSO_4 \cdot 5H_2O$	1.542	1.536	1.526	$Th(SO_4)_2 \cdot 8H_2O$[c]	1.560	1.544	1.530
$FeSO_4 \cdot 7H_2O$	1.4856	1.4782	1.4713	$Pu(SO_4)_2 \cdot 4H_2O$[c]	1.688	1.644	1.611
$FeOHSO_4 \cdot 2H_2O$	1.749	1.678	1.588	$TiOSO_4 \cdot 2H_2O$[f]	1.688	?	1.605

(continued)

Table 11.4 (continued)

Compounds	n_g	n_m	n_p	Compounds	n_g	n_m	n_p
$FeOHSO_4 \cdot 3H_2O$	1.621	1.598	1.516	$Ti(SO_4)_2 \cdot 4H_2O$[g]		1.653	1.500
$CoSO_4 \cdot H_2O$	1.645	?	1.600	$Zr(SO_4)_2 \cdot 4H_2O$[h]		1.655	1.614
$NaAl(SO_4)_2 \cdot 12H_2O$	1.463	1.461	1.449	$RbIn(SO_4)_2 \cdot 3H_2O$[i]	1.556	1.543	1.534
$NaNd(SO_4)_2 \cdot H_2O$[c]	1.620	1.578		$RbIn(SO_4)_2 \cdot 12H_2O$		1.4638	
$NaPu(SO_4)_2 \cdot H_2O$[c]	1.649	1.604		$RbV(SO_4)_2 \cdot 12\ H_2O$		1.469	
$Na_2Cu(SO_4)_2 \cdot 2H_2O$	1.601	1.578	1.544	$RbCr(SO_4)_2 \cdot 12H_2O$		1.4815	
$Na_2Mg(SO_4)_2 \cdot 4H_2O$	1.4869	1.4855	1.4826	$RbAl(SO_4)_2 \cdot 12H_2O$		1.4566	
$Na_2Ce(SO_4)_3 \cdot 3H_2O$[b]	1.698	?	1.650	$RbFe(SO_4)_2 \cdot 12H_2O$		1.4825	
$Na_2Co(SO_4)_2 \cdot 4H_2O$[j]	1.522	1.517	1.509	$RbRh(SO_4)_2 \cdot 12H_2O$		1.501	
$Na_2Ni(SO_4)_2 \cdot 4H_2O$[j]	1.525	1.520	1.514	$RbGa(SO_4)_2 \cdot 12H_2O$		1.4658	
$Na_2Mg(SO_4)_2 \cdot 4H_2O$[j]	1.487	1.485	1.481	$Rb_2Fe(SO_4)_2 \cdot 6H_2O$	1.4977	1.4874	1.4815
$Na_2Fe(SO_4)_2 \cdot 4H_2O$[j]	1.516	1.513	1.506	$Rb_2Mg(SO_4)_2 \cdot 6H_2O$	1.4779	1.4689	1.4672
$Na_2FeOH(SO_4)_4 \cdot 3H_2O$	1.586	1.525	1.508	$Rb_2Zn(SO_4)_2 \cdot 6H_2O$	1.4975	1.4884	1.4833
$Na_2Zn(SO_4)_2 \cdot 4H_2O$[j]	1.515	1.511	1.508	$Rb_2Mn(SO_4)_2 \cdot 6H_2O$	1.4907	1.4807	1.4767
$Na_2Mn(SO_4)_2 \cdot 4H_2O$[j]	1.566	1.543	1.531	$Rb_2Co(SO_4)_2 \cdot 6H_2O$	1.5014	1.4916	1.4859
$Na_2Ca_5(SO_4)_6 \cdot H_2O$	1.567	?	1.556	$Rb_2Cu(SO_4)_2 \cdot 6H_2O$	1.5036	1.4906	1.4886
$Na_2Ni(SO_4)_2 \cdot 4H_2O$[k]	1.520	1.518	1.513	$Rb_2Cd(SO_4)_2 \cdot 6H_2O$	1.4948	1.4848	1.4798
$Na_3Fe(SO_4)_3 \cdot 3H_2O$	1.614	1.558		$Rb_2Ni(SO_4)_2 \cdot 6H_2O$	1.5052	1.4961	1.4895
$Na_4Ca(SO_4)_3 \cdot 2H_2O$[l]		1.494	1.471	$Rb_4Pu(SO_4)_4 \cdot 2H_2O$[c]	1.563	1.539	1.534
$Na_4\ Mg_2(SO_4)_4 \cdot 5H_2O$		1.490		$Rb_4Ce(SO_4)_4 \cdot 2H_2O$[b]	1.591	1.564	1.552
$Na_4Fe_2(OH)_2(SO_4)_4 \cdot 3H_2O$	1.634	1.575	1.543	$Rb_4Ce(SO_4)_4 \cdot 3H_2O$[b]	1.594	1.567	1.555
$KAl(SO_4)_2 \cdot 12H_2O$		1.4565		$NH_4Al(SO_4)_2 \cdot 12H_2O$		1.4569	
$KGa(SO_4)_2 \cdot 12H_2O$		1.4653		$NH_4V(SO_4)_2 \cdot 12H_2O$		1.475	
$KCr(SO_4)_2 \cdot 12H_2O$		1.4814		$NH_4Cr(SO_4)_2 \cdot 12H_2O$		1.4842	

(continued)

Table 11.4 (continued)

Compounds	n_g	n_m	n_p	Compounds	n_g	n_m	n_p
$KFe(SO_4)_2 \cdot 12H_2O$		1.4817		$NH_4Fe(SO_4)_2 \cdot 12H_2O$		1.4848	
$KFe(SO_4)_2 \cdot 4H_2O^m$	1.629	1.602	1.582	$NH_4In(SO_4)_2 \cdot 12H_2O$		1.4664	
$KMgSO_4Cl \cdot 3H_2O$	1.516	1.505	1.494	$NH_4Rh(SO_4)_2 \cdot 12H_2O$		1.5103	
$KNd(SO_4)_2 \cdot 2H_2O^c$	1.571	1.556	1.545	$NH_4Ga(SO_4)_2 \cdot 12H_2O$		1.4684	
$KPu(SO_4)_2 \cdot 2H_2O^c$	1.587	1.573	1.562	$NH_4Nd(SO_4)_2 \cdot 4H_2O^n$	1.549	1.540	1.531
$KNd(SO_4)_2 \cdot H_2O^c$	1.602	1.579	1.553	$(NH_4)_2Zn(SO_4)_2 \cdot 6H_2O$	1.4994	1.4930	1.4888
$KPu(SO_4)_2 \cdot H_2O^c$	1.633	1.603	1.573	$(NH_4)_2Mg(SO_4)_2 \cdot 6H_2O$	1.4786	1.4730	1.4716
$K_2Mg(SO_4)_2 \cdot 2H_2O^o$	1.516		1.487	$(NH_4)_2Fe(SO_4)_2 \cdot 6H_2O$	1.4989	1.4915	1.4870
$K_2Mg(SO_4)_2 \cdot 4H_2O$	1.487	1.482	1.479	$(NH_4)_2\,Ca(SO_4)_2 \cdot H_2O$	1.536	1.532	1.524
$K_2Mg(SO_4)_2 \cdot 6H_2O$	1.4755	1.4629	1.4607	$(NH_4)_2Mn(SO_4)_2 \cdot H_2O$	1.4913	1.4840	1.4801
$K_2MgCa_2(SO_4)_4 \cdot 2H_2O$	1.567	1.560	1.547	$(NH_4)_2Co(SO_4)_2 \cdot 6H_2O$	1.5032	1.4953	1.4902
$K_2Ca(SO_4)_2 \cdot H_2O$	1.5176	1.5166	1.5010	$(NH_4)_2Cu(SO_4)_2 \cdot 6H_2O$	1.5054	1.5007	1.4910
$K_2Zn(SO_4)_2 \cdot 6H_2O$	1.4969	1.4833	1.4775	$(NH_4)_2Cd(SO_4)_2 \cdot 6H_2O$	1.4959	1.4887	1.4847
$K_2Fe(SO_4)_2 \cdot 6H_2O$	1.4969	1.4821	1.4759	$(NH_4)_2TiO(SO_4)_2 \cdot H_2O^p$		1.587	1.4949
$K_2Fe(SO_4)_2 \cdot 4H_2O^q$	1.509	1.501	1.497	$(NH_4)_2Ni(SO_4)_2 \cdot 6H_2O$	1.5081	1.5007	
$K_2Ni(SO_4)_2 \cdot 6H_2O$	1.5051	1.4916	1.4836	$(NH_4)_2Ca_5(SO_4)_6 \cdot H_2O$	1.595	1.580	1.567
$K_2Co(SO_4)_2 \cdot 6H_2O$	1.5004	1.4865	1.4807	$(NH_4)_4Ce(SO_4)_4 \cdot 2H_2O^b$	1.606	1.584	1.570
$K_2Cu(SO_4)_2 \cdot 6H_2O$	1.5020	1.4864	1.4836	$(NH_4)_4Pu(SO_4)_4 \cdot 2H_2O^c$	1.574	1.551	1.548
$K_2Ca_5(SO_4)_6 \cdot H_2O$	1.583	1.565	1.550	$(NH_4)_6Ce(SO_4)_5 \cdot 3H_2O^b$	1.579	?	1.540
$K_4Ce(SO_4)_4 \cdot 2H_2O^b$	1.594	1.557	1.549	$(NH_4)_8Ce(SO_4)_6 \cdot 5H_2O^b$	1.570	?	1.531
$K_4Ce(SO_4)_4 \cdot 3H_2O^b$	1.591	1.561	1.552	$CsV(SO_4)_2 \cdot 12H_2O$		1.478	
$K_4Pu(SO_4)_4 \cdot 2H_2O^c$	1.560	1.533	1.531	$CsCr(SO_4)_2 \cdot 12H_2O$		1.4810	
$K_4\,Cd_2(SO_4)_4 \cdot 3H_2O$		1.510		$CsAl(SO_4)_2 \cdot 12H_2O$		1.4586	
$K_4Mn_2(SO_4)_4 \cdot 3H_2O$		1.512		$CsGa(SO_4)_2 \cdot 12H_2O$		1.4650	
$CuFe_4(OH)_2(SO_4)_6 \cdot 20H_2O$	1.620	1.575	1.558	$CsIn(SO_4)_2 \cdot 12H_2O$		1.4652	

(continued)

Table 11.4 (continued)

Compounds	n_g	n_m	n_p	Compounds	n_g	n_m	n_p
$Cu_3Cd_2(OH)_6(SO_4)_2·4H_2O^r$	1.77	1.762	1.74	$CsTi(SO_4)_2·12H_2O$		1.4736	
$Cu_4(OH)_6SO_4·H_2O^s$	1.706	1.680	1.625	$CsFe(SO_4)_2·12H_2O$		1.4838	
$Cu_4(OH)_6SO_4·2H_2O^t$	1.694	1.682	1.637	$CsRh(SO_4)_2·12H_2O$		1.5077	
$Cu_4(OH)_{12}(SO_4)_2·3H_2O^u$	1.732	1.703	1.659	$Cs_2Mg(SO_4)_2·6H_2O$	1.4916	1.4858	1.4857
$Cu_6(OH)_{10}SO_4·H_2O^v$	1.723	1.721	1.693	$Cs_2Cd(SO_4)_2·6H_2O$	1.5062	1.5000	1.4795
$Cu_8(OH)_{10}(SO_4)_3·H_2O^w$	1.783	1.736	1.716	$Cs_2Zn(SO_4)_2·6H_2O$	1.5093	1.5048	1.5022
$MgAl(SO_4)_2F·18H_2O^x$	1.438	1.436	1.424	$Cs_2Mn(SO_4)_2·6H_2O$	1.5025	1.4966	1.4946
$MgAl_2(SO_4)_4·22 H_2O$	1.483	1.480	1.476	$Cs_2Cu(SO_4)_2·6H_2O$	1.5153	1.5061	1.5048
$MgFe_4(OH)_2(SO_4)_6·20H_2O$	1.575	1.535	1.510	$Cs_2Co(SO_4)_2·6H_2O$	1.5132	1.5085	1.5057
$Ca_3Mn(OH)_6(SO_4)_2·3H_2O^y$	1.682	1.656		$Cs_2Fe(SO_4)_2·6H_2O$	1.5094	1.5035	1.5003
$Ca_4Al_2(OH)_{12}SO_4·6H_2O$		1.504	1.488	$Cs_2Ni(SO_4)_2·6H_2O$	1.5162	1.5129	1.5087
$ZnFeOHSO_4·4H_2O^z$	1.688	1.640		$Tl_2Mn(SO_4)_2·6H_2O$	1.6084	1.5996	1.5861
$ZnFe_2(SO_4)_4·14H_2O^α$	1.578	1.568	1.522	$Tl_2Cu(SO_4)_2·6H_2O$	1.6190	1.6096	1.5996
$ZnAl_4(OH)_{12}SO_4·3H_2O^β$	1.527	1.525	1.517	$TlV(SO_4)_2·12H_2O$		1.514	
$Zn_4(OH)_6SO_4·5H_2O^γ$	1.567	1.565	1.532	$TlAl(SO_4)_2·12H_2O$		1.4975	
$AlOHSO_4·5H_2O^δ$	1.487	1.460	1.44	$Tl_2Zn(SO_4)_2·6 H_2O$	1.6168	1.6093	1.5931
$Al_2(OH)_4SO_4·3H_2O^ε$	1.545	1.525		$Tl_2Co(SO_4)_2·6H_2O$	1.6238	1.6176	1.6009
$Al_2OH(PO_4)SO_4·9H_2O^η$	1.499	1.484		$TlCr(SO_4)_2·12H_2O$		1.5228	
$Al_4(OH)_{10}SO_4·10H_2O$	1.471	1.471	1.463	$TlFe(SO_4)_2·12H_2O$		1.5237	
$AlFe(SO_4)_3·9H_2O^θ$	1.56	1.53		$TlRh(SO_4)_2·12H_2O$		1.548	
$VOSO_4·5H_2O^κ$	1.574	1.555	1.548	$TlGa(SO_4)_2·12H_2O$		1.5067	
$FeAl_2(SO_4)_4·22H_2O$	1.490	1.486	1.480	$Tl_2Mg(SO_4)_2·6H_2O$	1.5949	1.5884	1.5705
$FeOHSO_4·7H_2O^λ$	1.628	1.570	1.536	$Tl_2Fe(SO_4)_2·6H_2O$	1.6162	1.6093	1.5929
$Fe_5(OH)_2(SO_4)_6·20H_2O$	1.597	1.546	1.531	$Tl_2Ni(SO_4)_2·6H_2O$	1.6224	1.6183	1.6024

a[33], b[34], c[5], d[35], e[36], f[37], g[38], h[39], i[40], j[41], k[42], l[43], m[44], n[45], o[46], p[47], q[48], r[49], s[50], t[51], u[52], v[53], w[54], x[55], y[56], z[57], α[58], β[59], γ[60], δ[61], ε[62], η[63], θ[64], κ[65] and λ[66]

Table 11.5 Refractive indices of selenite hydrates

Compounds	n_g	n_m	n_p	Compounds	n_g	n_m	n_p
$BeSeO_4·4H_2O$	1.5027	1.5007	1.4664	$ZnSeO_4·6H_2O$[b]		1.5291	1.5039
$MgSeO_4·6H_2O$	1.4911	1.4892	1.4856	$CoSeO_4·6H_2O$	1.5227	1.5225	1.47
$CuSeO_4·H_2O$[a]	1.752	?	1.708	$NiSeO_4·6H_2O$[b]		1.5353	1.5125
$CuSeO_4·5\,H_2O$[a]	1.622	?	1.549				
$KAl(SeO_4)_2·12H_2O$[c]		1.4807		$Rb_2Mn(SeO_4)_2·6H_2O$	1.5258	1.5140	1.5094
$RbAl(SeO_4)_2·12H_2O$[c]		1.4810		$Cs_2Mn(SeO_4)_2·6H_2O$	1.5338	1.5279	1.5250
$NH_4Al(SeO_4)_2·12H_2O$[c]		1.4856		$Tl_2Mn(SeO_4)_2·6H_2O$	1.6531	1.6439	1.6276
$CsAl(SeO_4)_2·12H_2O$[c]		1.4865		$K_2Fe(SeO_4)_2·6H_2O$	1.5345	1.5182	1.5095
$TlAl(SeO_4)_2·12H_2O$[c]		1.5220		$(NH_4)_2Fe(SeO_4)_2·6H_2O$	1.5348	1.5280	1.5216
$(NH_4)_2Mg(SeO_4)_2·6H_2O$	1.5169	1.5093	1.5070	$Rb_2Fe(SeO_4)_2·6H_2O$	1.5328	1.5200	1.5133
$Rb_2Mg(SeO_4)_2·6H_2O$	1.5135	1.5031	1.5011	$RbFe(SeO_4)_2·12H_2O$		1.507	
$Cs_2Mg(SeO_4)_2·6H_2O$	1.5236	1.5179	1.5178	$Cs_2Fe(SeO_4)_2·6H_2O$	1.5414	1.5352	1.5306
$Tl_2Mg(SeO_4)_2·6H_2O$	1.6404	1.6337	1.6250	$CsFe(SeO_4)_2·12H_2O$		1.512	
$K_2Cu(SeO_4)_2·6H_2O$	1.5349	1.5228	1.5101	$Tl_2Fe(SeO_4)_2·6H_2O$	1.6589	1.6514	1.6352
$(NH_4)_2Cu(SeO_4)_2·6H_2O$	1.5387	1.5344	1.5201	$TlFe(SeO_4)_2·12H_2O$		1.524	
$Rb_2Cu(SeO_4)_2·6H_2O$	1.5318	1.5183	1.5153	$K_2Co(SeO_4)_2·6H_2O$	1.5380	1.5218	1.5158
$Cs_2Cu(SeO_4)_2·6H_2O$	1.5394	1.5298	1.5282	$Am_2Co(SeO_4)_2·6H_2O$	1.5417	1.5327	1.5261
$Tl_2Cu(SeO_4)_2·6H_2O$	1.6720	1.6565	1.6396	$Rb_2Co(SeO_4)_2·6H_2O$	1.5369	1.5256	1.5199
$K_2Zn(SeO_4)_2·6H_2O$	1.5335	1.5181	1.5121	$Cs_2Co(SeO_4)_2·6H_2O$	1.5453	1.5399	1.5354
$(NH_4)_2Zn(SeO_4)_2·6H_2O$	1.5385	1.5300	1.5240	$Tl_2Co(SeO_4)_2·6H_2O$	1.6590	1.6535	1.6442
$Rb_2Zn(SeO_4)_2·6H_2O$	1.5331	1.5222	1.5162	$K_2Ni(SeO_4)_2·6H_2O$	1.5427	1.5272	1.5181
$Cs_2Zn(SeO_4)_2·6H_2O$	1.5412	1.5362	1.5326	$(NH_4)_2Ni(SeO_4)_2·6H_2O$	1.5460	1.5370	1.5360
$Tl_2Zn(SeO_4)_2·6H_2O$	1.6615	1.6539	1.6414	$Rb_2Ni(SeO_4)_2·6H_2O$	1.5390	1.5291	1.5198
$(NH_4)_2Cd(SeO_4)_2·6H_2O$	1.5352	1.5260	1.5206	$Cs_2Ni(SeO_4)_2·6H_2O$	1.5489	1.5450	1.5395
$(NH_4)_2Mn(SeO_4)_2·6H_2O$	1.5288	1.5202	1.5160	$Tl_2Ni(SeO_4)_2·6H_2O$	1.6560	1.6498	1.6378

[a][67], [b][68] and [c][69]

Table 11.6 Refractive indices of phosphate hydrates and co-crystals with hydrogen peroxide

Compounds	n_g	n_m	n_p	Compounds	n_g	n_m	n_p
$HNaNH_4PO_4 \cdot 4H_2O$	1.469	1.442	1.439	$NH_4NiPO_4 \cdot 6H_2O$	1.534	1.530	1.522
$HMgPO_4 \cdot 3H_2O$	1.533	1.517	1.514	$(NH_4)_2Ca(HPO_4)_2 \cdot H_2O^a$	1.552	1.544	1.522
$HMgPO_4 \cdot 7H_2O$	1.486	1.485	1.477	$CuPbFeOH(PO_4)_2 \cdot H_2O^b$	2.00	1.93	1.90
$HCaPO_4 \cdot 2H_2O$	1.551	1.545	1.539	$CuFe_2(OH)_2(PO_4)_2 \cdot 4H_2O^c$	1.762	1.742	1.703
$HNa_2PO_4 \cdot 8H_2O^d$	1.458	1.457	1.443	$Cu_2Cd_2(SO_4)_5(PO_4)_2 \cdot H_2O^e$	1.669	1.636	1.624
$HAl_2Fe_3(PO_4)_4F \cdot 18H_2O^f$	1.534	1.531	1.522	$Be_3Ca(OH)_2(PO_4)_2 \cdot 4H_2O^g$	1.536	1.525	1.510
$H_2NaPO_4 \cdot 2H_2O$	1.4815	1.4629	1.4794	$Be_3Ca(OH)_2(PO_4)_2 \cdot 4H_2O^h$	1.530	1.520	1.520
$H_2KAl(PO_4)_2 \cdot H_2O$	1.539	1.536	1.522	$Be_3Ca_4Zn_2(PO_4)_6 \cdot 9H_2O^i$	1.580	1.560	1.556
$H_2KFe(PO_4)_2 \cdot H_2O$	1.630	1.614	1.592	$MgCaSc(PO_4)_2 \cdot 4H_2O^j$	1.582 ?		1.574
$H_2AlMnOH(PO_4)_2 \cdot H_2O^k$	1.544	1.529	1.511	$MgZr(PO_4)_2 \cdot 4H_2O^l$	1.635	1.622	1.597
$H_2Na_2Ca(PO_4)_2 \cdot 3H_2O^m$	1.506	1.504	1.496	$MgFe_2(OH)_2(PO_4)_2 \cdot 8H_2O^n$	1.670	1.637	1.584
$H_2Be_2Ca_3(PO_4)_4 \cdot 4H_2O^o$	1.586	1.566	1.560	$Mg_2OHPO_4 \cdot 3H_2O^p$	1.549	1.542	1.528
$H_2Mn_5(PO_4)_4 \cdot 4H_2O$	1.660	1.654	1.647	$Mg_3(PO_4)_2 \cdot 8 H_2O$	1.543	1.520	1.510
$H_4K_2Al_3(PO_4)_5 \cdot 11H_2O$	1.515	1.511	1.510	$Mg_3(PO_4)_2 \cdot 22H_2O$	1.469	1.465	1.461
$H_4(NH_4)_2Al_2(PO_4)_4 \cdot H_2O$	1.597	1.586	1.565	$CaZn_2(PO_4)_2 \cdot 2H_2O^q$	1.603	1.588	1.587
$H_4(NH_4)_2Fe_2(PO_4)_4 \cdot H_2O$	1.715	1.680	1.655	$Ca_2Mn_3O_2(PO_4)_3 \cdot 3H_2O^r$	1.83	1.81	1.79
$H_4Ca(PO_4)_2 \cdot H_2O$	1.529	1.515	1.496	$Ca_2Mn_4Fe_2(PO_4)_4 \cdot 2H_2O^s$	1.741	1.738	1.729
$H_6K_3Al_5(PO_4)_8 \cdot 13H_2O^t$	1.515	1.510		$Ca_4Al_2(PO_4)_2F_8 \cdot H_2O^u$	1.520	1.495	1.493
$H_7KCa_2(PO_4)_4 \cdot 2H_2O^v$	1.545	1.530	1.519	$Ca(H_2PO_4)_2 \cdot H_2O^w$	1.5292	1.5176	1.4932
$H_7NH_4Ca_2(PO_4)_4 \cdot 2H_2O^v$	1.553	1.536	1.525	$CaHPO_4 \cdot 2H_2O^w$	1.551	1.545	1.539
$H_8KFe_3(PO_4)_6 \cdot 6 H_2O$	1.601	1.595		$SrFe_3(OH)_6(PO_4)_2 \cdot 6H_2O^x$		1.872	1.862
$H_8NH_4Fe_3(PO_4)_6 \cdot 6H_2O$	1.591	1.580		$BaV_2O_2(PO_4)_2 \cdot 4H_2O^y$		1.721	1.715
$H_{10}K_2Al_6(PO_4)_{10} \cdot 15H_2O$	1.505	1.503	1.495	$BaV_3(OH)_6(PO_4)_2 \cdot 6H_2O^z$		1.858	1.817
$NaFe_9(OH)_{10}(PO_4)_6 \cdot 5H_2O^\alpha$	1.805	1.800	1.787	tr-$Zn_3(PO_4)_2 \cdot 4H_2O$	1.637	1.625	1.614

(continued)

Table 11.6 (continued)

Compounds	n_g	n_m	n_p	Compounds	n_g	n_m	n_p
NaMnOHPO$_4$·2H$_2$O[β]	1.737	1.715	1.699	o-Zn$_3$(PO$_4$)$_2$·4H$_2$O	1.599	1.598	1.589
Na$_2$Mg$_5$(PO$_4$)$_4$·7H$_2$O[γ]	1.543	1.540	1.538	YPO$_4$·2H$_2$O	1.645	1.612	1.605
Na$_2$HPO$_4$·H$_2$O	1.4873	1.4852	1.4556	CePO$_4$·H$_2$O	1.654	1.620	1.620
Na$_2$HPO$_4$·2H$_2$O	1.477	1.461	1.450	AlPO$_4$·2H$_2$O	1.593	1.588	1.565
Na$_2$HPO$_4$·7H$_2$O	1.4526	1.4424	1.4412	Al$_4$(OH)$_3$(PO$_4$)$_3$·9H$_2$O[δ]	1.578	1.560	1.511
Na$_2$HPO$_4$·12 H$_2$O	1.4373	1.4361	1.4321	Al$_4$Fe$_3$(OH)$_6$(PO$_4$)$_4$·2 H$_2$O[ε]	1.660	1.653	1.619
Na$_3$PO$_4$·H$_2$O	1.525	1.499		CrPO$_4$·6H$_2$O	1.599	1.591	1.568
Na$_3$PO$_4$·7H$_2$O	1.478	1.470	1.462	FePO$_4$·2H$_2$O	1.762	1.732	1.730
Na$_3$PO$_4$·8H$_2$O	1.471	1.468	1.458	FeZr(PO$_4$)$_2$·4H$_2$O[η]	1.650	1.650	1.646
Na$_3$PO$_4$·12H$_2$O	1.4524	1.4458		Fe$_3$(PO$_4$)$_2$·8H$_2$O	1.6294	1.6024	1.5788
Na$_7$(PO$_4$)$_2$F·19H$_2$O		1.4519		Fe$_3$(OH)$_2$(PO$_4$)$_2$·8H$_2$O[θ]	1.723	1.682	1.628
KH$_2$PO$_4$·3H$_2$O	1.480	1.477	1.468	Fe$_3$(OH)$_3$(PO$_4$)$_2$·5H$_2$O[κ]	1.747	1.675	1.662
KAl$_2$OH(PO$_4$)$_2$·2H$_2$O	1.568	1.564	1.562	Fe$_5$(OH)$_2$(PO$_4$)$_4$·2H$_2$O[λ]	1.812	1.803	1.775
KFe$_2$OH(PO$_4$)$_2$·2H$_2$O	1.741	1.720	1.706	Co$_3$(PO$_4$)$_2$·8H$_2$O[ν]	1.631	1.600	1.581
KAl$_2$OH(PO$_4$)$_2$·2H$_2$O[μ]	1.604	1.597	1.591	Ni$_3$(PO$_4$)$_2$·8H$_2$O[π]	1.68	?	1.63
KFe$_7$(OH)$_7$(PO$_4$)$_5$·8H$_2$O[ξ]	1.800	1.785	1.780	Na$_3$PO$_4$·4H$_2$O$_2$·2H$_2$O	1.487	1.455	1.440
NH$_4$MgPO$_4$·6H$_2$O	1.504	1.496	1.495	Na$_2$HPO$_4$·1.5H$_2$O[ρ]	1.523	1.493	1.475
NH$_4$CoPO$_4$·6H$_2$O	1.535	1.530	1.525	Na$_2$HPO$_4$·3.5H$_2$O[ρ]	1.488	1.449	1.442

a[69], cit. in [70], b[71], c[72], d[73], e[74], f[75], g[76], h[77], i[78], j[79], k[80], l[81], m[82], n[83], o[84], p[85], q[86], r[87], s[88], t[89], u[90], v[91], w[92], x[93], y[94], z[95], α[96], β[97], γ[98], δ[99], ε[100], η[101], θ[102], κ[103], λ[104], μ[105], ν[106], ξ[107], π[108] and ρ[109]

Table 11.7 Refractive indices of arsenate hydrates

Compounds	n_g	n_m	n_p	Compounds	n_g	n_m	n_p
HNaNH$_4$AsO$_4$·4H$_2$O	1.4791	1.4663	1.4649	Mg$_3$(AsO$_4$)$_2$·8H$_2$O	1.596	1.571	1.563
HMg(AsO$_4$)$_4$·4H$_2$O[a]	1.562	1.546	1.531	CaZn$_5$Fe$_2$(AsO$_4$)$_6$·14H$_2$O[b]	1.656	1.631	1.628
HMgAsO$_4$·7H$_2$O	1.550	1.53	1.525	Ca$_2$Zn(AsO$_4$)$_2$·2H$_2$O[c]	1.720	1.710	1.707
HCaAsO$_4$· H$_2$O	1.638	1.602	1.590	Ca$_2$Mn(AsO$_4$)$_2$·2H$_2$O[d]	1.751	1.721	1.701
HCaAsO$_4$·2 H$_2$O	1.594	1.589	1.583	Ca$_2$Fe$_3$O$_2$(AsO$_4$)$_2$·2H$_2$O[e]	1.933	1.923	1.810
HCaAsO$_4$·3H$_2$O	1.532	1.524	1.513	Ca$_2$Ni(AsO$_4$)$_2$·2H$_2$O[f]	1.735	1.720	1.715
HMnAsO$_4$·4H$_2$O[g]	1.686	1.639	1.620	Ca$_3$(AsO$_4$)$_2$·9 H$_2$O[h]	1.593	?	1.585
H$_2$NaAsO$_4$·H$_2$O	1.5607	1.5535	1.5382	CaMn(HAsO$_4$)$_2$·2H$_2$O[i]	1.642	1.624	1.618
H$_2$NaAsO$_4$·2H$_2$O	1.5265	1.5021	1.4794	BaFe$_4$(OH)$_5$(AsO$_4$)$_3$·5H$_2$O[j]		1.728	1.718
H$_2$Ca$_5$(AsO$_4$)$_4$·5H$_2$O	1.615	1.614	1.613	ZnFe$_2$(OH)$_2$(AsO$_4$)$_2$·4H$_2$O[k]	1.798	1.730	1.696
H$_2$Ca$_5$(AsO$_4$)$_4$·9H$_2$O[l]	1.585	1.572	1.562	Zn$_2$Pb(AsO$_4$)$_2$·2H$_2$O[m]	1.98	1.80	1.72
NaAl$_4$(OH)$_4$(AsO$_4$)$_3$·4H$_2$O[n]		1.556		Zn$_2$Fe$_3$ (OH)$_2$(AsO$_4$)$_2$·4H$_2$O[k]	1.712	1.678	1.672
NaCu$_5$Ca(AsO$_4$)$_4$Cl·5H$_2$O[o]	1.749	1.675		Zn$_3$(AsO$_4$)$_2$·8H$_2$O	1.671	1.638	1.622
Na$_3$AsO$_4$·12H$_2$O	1.4669	1.4589		Al$_2$(OH)$_3$AsO$_4$·3H$_2$O[p]	1.548	1.544	1.540
Na$_7$(AsO$_4$)$_2$F·19H$_2$O		1.4693		Pb$_3$Sb(OH)$_6$SO$_4$AsO$_4$·3H$_2$O[q]	1.801	1.760	
Na$_2$HAsO$_4$·7H$_2$O	1.4782	1.4658	1.4622	BiNi$_2$OH(AsO$_4$)$_2$·H$_2$O[r]	1.97	1.95	1.92
Na$_2$HAsO$_4$·12 H$_2$O	1.4513	1.4496	1.4453	Mn$_3$(AsO$_4$)$_2$·4H$_2$O[s]	1.671	?	1.656
KAl$_4$(OH)$_4$(AsO$_4$)$_3$·6H$_2$O[t]		1.565		FeAsO$_4$·2H$_2$O	1.814	1.795	1.784
CuCaFeOH(AsO$_4$)$_2$·H$_2$O[u]	1.89	1.834	1.830	Fe$_3$(AsO$_4$)$_2$·8H$_2$O	1.702	1.668	1.635
Cu$_2$Al$_2$(OH)$_4$(AsO$_4$)$_2$·H$_2$O[v]	1.796	1.773	1.752	Fe(OH)$_3$(AsO$_4$)$_2$·3H$_2$O[w]	1.835	1.83	1.825
Cu$_3$(AsO$_4$)$_2$·4 H$_2$O[x]	1.760	1.755	1.745	Co$_3$(AsO$_4$)$_2$·8H$_2$O	1.699	1.661	1.626
MgCa$_2$(AsO$_4$)$_2$·H$_2$O[y]	1.713	1.703	1.694	Ni$_3$(AsO$_4$)$_2$·8H$_2$O	1.687	1.658	1.622
MgCa(OH)$_5$AsO$_4$·5H$_2$O[z]	1.563	1.548	1.540				

[a][110], [b][111], [c][112], [d][113], [e][114], [f][115], [g][116, [h][117], [i][118], [j][119], [k][120], [l][121], [m][122], [n][123], [o][124], [p][125], [q][126], [r][127], [s][128], [t][129], [u][130], [v][131], [w][132], [x][133], [y][134] and [z][135]

Table 11.8 Refractive indices of halogenate hydrates

Compounds	n_g	n_m	n_p	Compounds	n_g	n_m	n_p
$LiClO_4 \cdot 3H_2O$		1.483	1.448	Sm $(ClO_4)_3 \cdot 4Q \cdot 9H_2O^a$	1.473	1.464	1.460
$Cu(ClO_4)_2 \cdot 6H_2O$	1.522	1.505	1.495	$Mg(BrO_3)_2 \cdot 6H_2O$		1.5139	
$Mg(ClO_4)_2 \cdot 6H_2O$		1.482	1.458	$Zn(BrO_3)_2 \cdot 6H_2O$		1.5452	
$Ba(ClO_3)_2 \cdot H_2O$	1.635	1.577	1.562	$La(BrO_3)_3 \cdot 9H_2O^b$	1.595	?	1.544
$Ba(ClO_4)_2 \cdot 3H_2O$	1.5330	1.5323		$Pr(BrO_3)_3 \cdot 9H_2O^b$	1.598	?	1.546
$Zn(ClO_4)_2 \cdot 6H_2O$		1.508	1.487	$Nd(BrO_3)_3 \cdot 9H_2O^b$	1.599	?	1.547
$Cd(ClO_4)_2 \cdot 6H_2O$		1.489	1.480	$Sm(BrO_3)_3 \cdot 9H_2O^b$	1.601	?	1.549
$Hg(ClO_4)_2 \cdot 6H_2O$		1.511	1.509	$Eu(BrO_3)_3 \cdot 9H_2O^b$	1.602	?	1.551
$Mn(ClO_4)_2 \cdot 6H_2O$		1.492	1.475	$Gd(BrO_3)_3 \cdot 9H_2O^b$	1.603	?	1.551
$Fe(ClO_4)_2 \cdot 6H_2O$		1.493	1.478	$Tb(BrO_3)_3 \cdot 9H_2O^b$	1.605	?	1.552
$Co(ClO_4)_2 \cdot 6H_2O$		1.510	1.490	$Dy(BrO_3)_3 \cdot 9H_2O^b$	1.605	?	1.553
$Ni(ClO_4)_2 \cdot 6H_2O$		1.518	1.498	$Er(BrO_3)_3 \cdot 9H_2O^b$	1.607	?	1.555
La $(ClO_4)_3 \cdot 4Q \cdot 9H_2O^a$	1.464	1.457	1.454	$3Cu(IO_3)_2 \cdot 2H_2O$	1.99	1.90	1.890
Ce $(ClO_4)_3 \cdot 4Q \cdot 9H_2O^a$	1.465	1.458	1.455	$Ca(IO_3)_2 \cdot 6H_2O$	1.686	1.644	1.604
Pr $(ClO_4)_3 \cdot 4Q \cdot 9H_2O^a$	1.464	1.456	1.453	$Ca(IO_4)_2 \cdot 6H_2O$	1.686	1.644	1.604
Nd $(ClO_4)_3 \cdot 4Q \cdot 9H_2O^a$	1.469	1.461	1.458				

^{a}Q = dioxane $C_4H_8O_2$ [136] and ^{b}[137]

References

1. E.A. Ingerson, G.W. Morey, Am. Miner. **36**, 778 (1951)
2. R. Hodenberg, G. Struensee, Ns. Jb. Miner. Monatsh. **3**, 125 (1980)
3. A.V. Novoselova, A.S. Pashinkin, K.N. Semenenko, Bull. Moscow Univ. (in Russian) **3**, 49 (1955)
4. N.K. Bel'skiy, Dokl. Acad. Sci. USSR (in Russian) **143**, 1313 (1962)
5. E. Staritzky, A.L. Truitt, in *The actinide elements*, ed. by G.T. Seaborg, J.J. Katz, (McGraw-Hill, New York, 1954)
6. G.B. Bokii, T.S. Khodasheva, Crystallography (in Russian) **1**, 197 (1956)
7. R.A. Chupakhina, Yu.V. Indukaev, V.V. Serebrennikov, Rus. J. Inorg. Chem. (in Russian) **6**, 2713 (1961)
8. A.N. Sergeeva, K.N. Mikhalevich, Rus. J. Inorg. Chem. (in Russian) **7**, 686 (1962)
9. O.M. Ansheles, T.N. Burlakova, *Crystallooptics as foundation of microanalysis*, (Leningrad Univ. Press, 1948)
10. L.N. Komissarova, G.Ya. Pushkina, V.I. Spitsyn, Rus. J. Inorg. Chem. (in Russian) **8**, 1384 (1963)
11. A. Jayaraman, Proc. India Acad. Sci. A **47**, 147 (1958)
12. E. Staritzky, Analyt. Chem. **28**, 2022 (1956)
13. E. Staritzky, Analyt. Chem. **28**, 2021 (1956)
14. A.C. Roberts, H.G. Ansell, I.R. Jonasson et al., Can. Miner. **24**, 51 (1986)
15. J. Suzuki, M. Ito, J. Jap. Assoc. Miner. **68**, 353 (1973)

16. A.P. Sabina, J.L. Jambor, A.G. Plant, Can. Miner. **9**, 468 (1968)
17. N.A. Moshkina, I.A. Poroshina, Proc. Sib. Div. Acad. Sci. USSR, Chem. (in Russian) **5**, 122 (1977)
18. K. Nagashima, R. Miyawaki, J. Takase et al., Am. Miner. **71**, 1028 (1986)
19. V.N. Yakovenchuk, Ya.A. Pakhomovsky, A.V. Voloshina et al., Miner. Journ. (in Russian) **12**, 74 (1990)
20. A.P. Khomyakov, N.G. Shumyatskaya, L.I. Polezhaeva, Proc. Rus. Miner. Soc. **6**, 129 (1992)
21. V.N. Yakovenchuk, Yu.P. Men'shikov, Ya.A. Pakhomovsky et al., Proc. Rus. Miner. Soc. (in Russian) **1**, 96 (1997)
22. J.D. Grice, R.A. Gault, A.C. Roberts et al., Can. Miner. **38**, 1457 (2000)
23. J.D. Grice, R.A. Gault, J. Van Velthuizen, Can. Miner. **35**, 181 (1997)
24. W.D. Birch, U. Kolitsch, T. Witzke et al., Can. Miner. **38**, 1467 (2000)
25. I.V. Pekov, N.V. Chukanov, N.V. Zubkova et al., Can. Miner. **48**, 95 (2010)
26. J.L. Jambor, D.G. Fong, A.P. Sabina, Can. Miner. **10**, 84 (1969)
27. N.P. Tomilov, A.G. Merkulov, A.S. Berger, Rus. J. Inorg. Chem. (in Russian) **14**, 3000 (1969)
28. N.P. Tomilov, A.S. Berger, I.A. Vorsina et al., Proc. Sib. Div. Acad. Sci. USSR, Chem. (in Russian), No 4, 87 (1970)
29. N.P. Tomilov, A.S. Berger, I.A. Poroshina, Proc. Sib. Div. Acad. Sci. USSR, Chem. (in Russian), No 14, 32 (1970)
30. J.L. Jambor, A.P. Sabina, B.D. Sturman, Can. Miner. **15**, 399 (1977)
31. R.A. Gault, G.Y. Chao, Can. Miner. **35**, 1541 (1997)
32. N.P. Luzhnaya, I.S. Kovaleva, Rus. J. Inorg. Chem. (in Russian) **6**, 1440 (1961)
33. I.G. Druzhinin, O.A. Kosyakina, Rus. J. Inorg. Chem. (in Russian) **6**, 1702 (1961)
34. V.A. Golovnya, L.A. Pospelova, Rus. J. Inorg. Chem. (in Russian) **6**, 1574 (1961)
35. E.N. Deichman, Rus. J. Inorg. Chem. (in Russian) **6**, 1671 (1961)
36. V.A. Nikolaev, N.K. Razumovsky, Proc. Miner. Soc. USSR (in Russian) **62**, 150 (1933)
37. B.E. Boguslavskaya, O.M. Ottamanovskaya, Rus. J. Gen. Chem. (in Russian) **10**, 673 (1940)
38. B.E. Boguslavskaya, Rus. J. Gen. Chem. (in Russian) **9**, 1084 (1939)
39. E. Staritzky, J. Singer, Analyt. Chem. **28**, 553 (1956)
40. E.N. Deichman, G.V. Rodicheva, Rus. J. Inorg. Chem. (in Russian) **6**, 2180 (1961)
41. G.B. Bokii, Bull. Moscow Univ. (in Russian) **10**, 175 (1948)
42. E.H. Nickel, P.J. Bridge, Miner. Mag. **42**, 37 (1977)
43. L. Vergouven, Am. Miner. **66**, 632 (1981)
44. A. Kozenzweig, E.B. Gross, Am. Miner. **40**, 469 (1955)
45. E. Staritzky, D.T. Cromer, Analyt. Chem. **28**, 554 (1956)
46. I.S. Rassonskaya, Rus. J. Inorg. Chem. (in Russian) **1**, 1284 (1956)
47. D.L. Motov, Rus. J. Inorg. Chem. (in Russian) **2**, 2661 (1957)
48. G. Giester, B. Rieck, Eur. J. Miner. **7**, 559 (1995)
49. P. Elliot, J. Brugger, T. Caradoc-Davies, Miner. Mag. **74**, 39 (2010)
50. A.I. Komkov, E.I. Nefedov, Proc. Miner. Soc. USSR (in Russian) **96**, 58 (1967)
51. P.J. Dunn, R.C. Rouse, J.A. Nelen, Miner. Mag. **40**, 1 (1975)
52. O. Medenbach, W. Gebert, Ns. Jb. Min. Monatsh. **3**, 401 (1993)
53. J.J. Pluth, I.M. Steele, A.R. Kampf et al., Miner. Mag. **69**, 973 (2005)
54. G.A. Yurgenson, N.G. Smirnova, L.A. Karenina, Bull. Geograf. Soc. USSR (in Russian), **9**, 3(1968)
55. S.A. Williams, F.P. Cesbron, Miner Mag. **47**, 37 (1983)
56. C. Gaudefroy, M.M. Grander, F. Permingeat et al., Bull. Soc. France. Miner. Crist. **91**, 43 (1968)
57. W. Li, G. Chen, Z. Peng, Am. Miner. **73**, 1493 (1988)
58. W. Li, G. Chen, Acta Miner. Sinica **10**, 299 (1990)
59. P. Haynes, J. Grice, H.T. Evans, Can. Miner. **41**, 83 (2003)
60. M. Ohnishi, I. Kusachi, S. Kobayashi, Can. Miner. **45**, 1511 (2007)

61. P. Bariand, J.P. Berthelon, F.P. Cesbron et al., Comp. Rend. D **277**, 1585 (1973)
62. G.M. Gamyanin, Yu.Ya. Zhdanov, N.V. Zayakina et al., Proc. Rus. Miner. Soc. (in Russian) **4**, 20 (2006)
63. M.E.J. de Abeledo, V. Angelelli, M.A.R. de Benyacar, Amer. Miner. **53**, 1 (1968)
64. F. Demartin, C. Castellano, C.M. Gramaccioli et al., Can. Miner. **48**, 1465 (2010)
65. M.A. Cooper, F.C. Hawthorne, J.D. Grice et al., Can. Miner. **41**, 959 (2003)
66. L. Hilin, Z. Jingliang, L. Jiaju, Am. Miner. **69**, 1194 (1984)
67. N.M. Selivanova, A.I. Maier, K.K. Samplavskaya, Rus. J. Inorg. Chem. (in Russian) **7**, 1074 (1962)
68. Landolt-Börnstein, Band II, Teil 8. *Optische Konstanten*, eds. K.H. Hellwege, A.M. Hellwege (Springer, Berlin, 1962)
69. K. Wendekamm, Z. Krist. **85**, 169 (1933)
70. J.E.J. Martini, Bull. South African Speleological Assoc. **32**, 72 (1991), cit. in J.L. Jambor, J. Puziewicz, Am. Miner. **78**, 1108 (1993)
71. W. Krause, K. Belendorff, H.-J. Bernhardt et al., Ns. Jb. Min. Monatsh. **2**, 111 (1998)
72. S.J. Mills, U. Kolitsch, W.D. Birch et al., Aust. J. Miner. **14**, 3 (2008)
73. T.V. Ruiz, R.J. Sureda, Can. Miner. **42**, 1906 (2004)
74. P. Elliott, J. Brugger, A. Pring et al., Am. Miner. **93**, 910 (2008)
75. J.P. Richardson, A.C. Roberts, J.D. Grice, Can. Miner. **26**, 971 (1988)
76. N.A. Grigor'ev, Proc. Miner. Soc. USSR (in Russian) **93**, 156 (1964)
77. F. Walter, Eur. J. Miner. **4**, 1275 (1992)
78. G.W. Robinson, J.D. Grice, J. Van Velthuizen, Can. Miner. **23**, 507 (1985)
79. R.P. Liferovich, V.N. Yakovenchuk, Ya.F. Pakhomovsky et al., Proc. Rus. Miner. Soc. (in Russian) **4**, 80 (1997)
80. D.R. Peacor, P.J. Dunn, W.L. Roberts et al., Am. Miner. **69**, 380 (1984)
81. F.C. Hawthorne, N.A. Ball, J.W. Nizamoff et al., Miner. Mag. **73**, 415 (2009)
82. D.R. Peacor, P.J. Dunn, W.B. Simmons et al., Miner. Records **16**, 467 (1985)
83. B.V. Chesnokov, V.A. Vilisov, G.E. Cherepinskaya et al., Proc. Miner. Soc. USSR (in Russian) **1**, 42 (1983)
84. D.R. Peacor, P.J. Dunn, W.L. Roberts et al., Bull. Minér. **106**, 499 (1983)
85. Yu.L. Kapustin, A.V. Bykona, Z.V. Pudovkina, Proc. Miner. Soc. USSR (in Russian) **3**, 341 (1980)
86. B.D. Sturman, R.C. Rouse, P.J. Dunn, Am. Miner. **66**, 843 (1981)
87. A.C. Roberts, B.D. Sturman, P.J. Dunn, Can. Miner. **27**, 451 (1989)
88. M.A. Galliske, M.A. Cooper, F.C. Hawthorne et al., Am. Miner. **84**, 1674 (1999)
89. F. Balenzano, L. Dell' Anna, M. Di Pierro, Ns. Jb. Min. Monatsh. **2**, 49 (1976)
90. A.R. Kampf, F. Colombo, W.B. Simmons et al., Am. Miner. **95**, 392 (2010)
91. W.E. Brown, J.P. Smith, J.R. Lehr, A.W. Frazier, J. Phys. Chem. **62**, 625 (1958)
92. R.W. Mooney, M.A. Aia, Chem. Rev. **61**, 433 (1961)
93. K. Walenta, W.D. Birch, P.J. Dunn, Chem. Erde.-Geochem. **56**, 171 (1996)
94. A. Pring, U. Kolitsch, W.D. Birch et al., Miner. Mag. **63**, 735 (1999)
95. U. Kolitsch, M.R. Taylor, G.D. Fallon et al., Am. Miner. **85**, 1324 (2000)
96. P.B. Moore, J. Ito, Miner. Mag. **42**, 137 (1978)
97. A.-M. Fransolet, M.A. Cooper, P. Černy et al., Can. Miner. **38**, 893 (2000)
98. O.V. Yakubovich, W. Massa, R.P. Liferovich et al., Can. Miner. **38**, 831 (2000)
99. A.-M. Fransolet, Bull. Minér. **110**, 647 (1987)
100. B.D. Sturman, J.A. Mandarino, M.E. Mrose et al., Can. Miner. **19**, 381 (1981)
101. C. Milton, J.J. McGee, H.T. Evans, Am. Miner. **78**, 437 (1993)
102. D.R. Peacor, P.J. Dunn, W.B. Simmons, Ns. Jb. Min. Monatsh. **11**, 524 (1983)
103. U. Kolitsch, H.-J. Bernhardt, C.L. Lengauer et al., Eur. J. Miner. **18**, 793 (2006)
104. P. Keller, Ns. Jb. Min. Monatsh. **2**, 49 (1980)
105. P.J. Dunn, R.C. Rouse, T.J. Campbell et al., Amer. Miner. **69**, 374 (1984)
106. V.N. Yakovenchuk, G.Y. Ivanuk, Y.A. Mikhailova et al., Can. Miner. **44**, 117 (2006)
107. W.D. Birch, A. Pring, P.G. Self et al., Miner. Mag. **60**, 787 (1996)

108. V.E. Buchwald, Ns. Jb. Min. Monatsh. **2**, 76 (1990)
109. E.D. Ruchkin, E.A. Ukraintseva, J. Struct. Chem. **4**, 850 (1963)
110. F. Fontan, M. Orliac, F. Permingeat et al., Bull. Soc. Frans. Minér. Crist. **96**, 365 (1973)
111. O. Medenbach, K. Schmetzer, K. Abraham, Ns. Jb. Min. Monatsh. **4**, 167 (1988)
112. P. Keller, J. Innes, P.J. Dunn, Ns. Jb. Miner. Abh. **156**, No 11, 523 (1986)
113. P.J. Dunn, D.R. Peacor, Su Shu-Chun et al., Ns. Jb. Miner. Abh. **157**, No 2, 113 (1987)
114. A.V. Voloshin, Yu.P. Men'shikov, L.I. Polezhaeva et al., Miner. J. (in Russian) **4**, 90(1982)
115. N.V. Chukanov, A.A. Mukhanova, Sh. Mekkel et al., Proc. Rus. Miner. Soc. (in Russian) **4**, 32 (2009)
116. F. Fontan, M. Orliac, F. Permingeat, Bull. Soc. Frans. Minér. Crist. **98**, 78 (1975)
117. K. Walenta, Tscherm. Miner. Petrogr. Mitt. **24**, 125 (1977)
118. H. Bari, F.P. Cesbron, F. Permingeat et al., Bull. Minér. **103**, 122 (1980)
119. K. Walenta, W. Wimmenauer, Tscherm. Miner. Petrogr. Mitt. **11**, 121 (1966)
120. F.P. Cesbron, M. Romero, S.A. Williams, Bull. Minér. **104**, 582 (1981)
121. H. Bari, F. Permingeat, R. Pierrot et al., Bull. Minér. **103**, 541 (1980)
122. P. Süsse, G. Schnorrer, Ns. Jb. Miner. Monatsh. **3**, 118 (1980)
123. M.S. Rumsey, S.J. Mills, J. Spratt, Miner. Mag. **74**, 929 (2010)
124. R. Ondruš, F. Veselovský, R. Skála et al., Can. Miner. **44**, 523 (2006)
125. K. Walenta, Aufschluss. **34**, 445 (1983)
126. I. Sima, Can. Miner. **41**, 1314 (2003)
127. W. Krause, H.-J. Bernhardt, H. Effenberger et al., Eur. J. Miner. **14**, 115 (2002)
128. P.J. Dunn, Amer. Miner. **66**, 182 (1981)
129. K. Schmetzer, W. Horn, H. Bank, Ns. Jb. Miner. Monatsh. **3**, 97 (1981)
130. W. Krause, G. Blass, H.-J. Bernhardt et al., Can. Miner. **40**, 1223 (2002)
131. S.A. Williams, Miner. Mag. **41**, 27 (1977)
132. N.V. Chukanov, I.V. Pekov, Sh. Mekkel et al., Proc. Rus. Miner. Soc. (in Russian) **3**, 100 (2009)
133. H. Sarp, R. Cerny, Eur. J. Miner. **12**, 1045 (2000)
134. P.J. Dunn, B.D. Sturman, J.A. Nelen, Am. Miner. **72**, 217 (1987)
135. K. Walenta, P.J. Dunn et al., Aufschluss. **40**, 369 (1989)
136. W.G.R. de Camargo, J.V. Valarelli, Acta Cryst. **16**, 321 (1963)
137. H. Schumann, Z. anorg. allgem. Chem. **271**, 29 (1952)

Part IV
Refractive Indices of Selected Organic Compounds

Chapter 12
Refractive Indices of Selected Organic Compounds

This section lists the RIs of selected monomeric and polymeric organic compounds of particular importance (see also Sect. 3.3). Polymeric materials with high RI, low birefringence (Δn) and good optical transparency have received much attention due to their wide range of technological applications. Compared to inorganic glasses, they have lighter weight, better impact resistance, processability and dying ability, as well as lower cost. These materials are highly required for advanced optoelectronic applications including optical encapsulants or adhesives for anti-reflective coatings (Table 12.1).

Liquid crystals, which occupy an intermediate place between crystalline and liquid substances, have extremely important technological applications (especially display devices) which specially require high birefringenices. Table 12.2 contains the data on a series of linear para-arene-alkyne derivatives, studied as a guest-host nematic mixture.

Numerous examples of polymeric compounds with high RIs, which fall outside the scope of this book, are reviewed in Refs. [17–22].

© The Author(s) 2016
S.S. Batsanov et al., *Refractive Indices of Solids*,
SpringerBriefs in Applied Sciences and Technology,
DOI 10.1007/978-981-10-0797-2_12

Table 12.1 Refractive indices of organic and organometallic solids

Compounds	n_g	n_m	n_p	Compounds	n_g	n_m	n_p
$CO(NH_2)_2$[a]	1.603	1.484		$Ce_2(C_2O_4)_3\cdot 11H_2O$[j]	1.610	1.551	1.478
$CS(NH_2)_2$[b]	1.797	1.783	1.628	$Nd_2(C_2O_4)_3\cdot 11H_2O$[j]	1.612	1.552	1.482
$CSe(NH_2)_2$[b]	1.837	1.793	1.788	$Eu_2(C_2O_4)_3\cdot 11H_2O$[j]	1.615	1.554	1.485
CH_3CONH_2[c]		1.495	1.460	$Pu_2(C_2O_4)_3\cdot 11H_2O$[j]	1.636	1.579	1.502
$SO(CH_3)_2$[d]		1.4772		Teflon, $(CF_2)_n$		1.315	
$KH_5C_8O_4$[e]		1.6666		$C_3H_6N_6\cdot HPO_3$[k]		1.640	
$Na_2C_2O_4$[f]	1.592	1.524	1.415	$C_3H_6N_6\cdot H_3PO_4$[k]	1.725	?	1.475
MgC_2O_4[g]	1.595	1.530	1.365	$(C_5O_2H_8)_n$[l]		1.4925	
$CaC_2O_4\cdot 3\,H_2O$[h]	1.533	1.516	1.483	$(C_6H_9NO)_n$[m]		1.5274	
$Ca(CHO_2)_2$	1.578	1.514	1.510	$(C_6H_{10}O_5)_n$[n]		1.4701	
$SrC_2O_4\cdot 2\,H_2O$	1.535	1.517		$C_6H_{13}NO$		1.528	
$Sr(CHO_2)_2$	1.598	1.574	1.559	$(C_8H_8)_n$[o]		1.5916	
$Ba(CHO_2)_2$	1.636	1.597	1.575	C_9H_8O[p]		1.6209	
$Zn(CHO_2)_2\cdot 2H_2O$	1.566	1.526	1.513	$C_{10}H_8$[q]		1.5821	
CdC_2O_4	1.570	1.558	1.555	$C_{14}H_{10}$[r]		1.5948	
$Th(C_2O_4)_2$	1.556	?	1.471	$(C_{16}H_{14}O_3)_n$[s]		1.5848	
PbC_2O_4	1.740	1.722	1.715	$(C_{34}H_{20}N_4O_2S_3)_n$[t]	1.7606	1.7530	
$MnC_2O_4\cdot 2H_2O$[i]	1.650	1.550	1.424	$(C_{33}H_{19}N_5O_2S_3)_n$[t]	1.7638	1.7564	
$La_2(C_2O_4)_3\cdot 10H_2O$	1.597	1.543	1.472	$(C_{40}H_{24}N_4O_2S_5)_n$[t]	1.7718	1.7668	

(a) [1], (b) [2], (c) [3], (d) [4], (e) [5], (f) [6], (g) [7], (h) [8], (i) [9], (j) [10], (k) [11], (l) polymethyl methacrylate, PMMA [12], (m) polyvinyl pyrrolidone, PVP [13], (n) cellulose [14], (o) polystyrene [14], (p) [15], (q) naphtalene, (r) anthracene, (s) polycarbonate [14], (t) polyamides [15]

Table 12.2 Refractive indices of selected liquid crystalline materials $H_{2x+1}C_x$–X-C_6H_4-C≡C-C_6H_3(R)-C≡C-C_6H_4-X-C_xH_{2x+1} (λ = 589 nm) [16]

x	R	X	n_g	n_m
3	Me	–	2.05	1.56
4	Me	–	2.01	1.59
5	Me	–	2.02	1.59
6	Me	–	1.92	1.55
7	Me	–	1.94	1.56
5	Et	–	1.92	1.55
5	Me	O	2.12	1.61
5	Me	S	2.21	1.60

References

1. P.J. Bridge, Miner. Mag. **39**, 346 (1973)
2. V.F. Dvoryankin, E.D. Ruchkin, J. Struct. Chem. **3**, 325 (1962)
3. B.I. Srebrodol'sky, Proc. Miner. Soc. USSR (in Russian) **104**, 326 (1975)
4. I.Z. Kozma, P. Krok, E. Riedle, J. Opt. Soc. Am. **B22**, 1479 (2005)
5. K. Moutzouris, I. Stavrakas, D. Triantis, M. Enculescu, Optics Mater. **33**, 812 (2011)

6. A.P. Khomyakov, Proc. Rus. Miner. Soc. (in Russian) **1**, 126 (1996)
7. Yu.A. Zhemchuzhnikov, A.I. Ginsburg, *Foundations of the petrology of coals* (in Russian). Acad. Sci. USSR, 93 (1960)
8. R. Basso, G. Lucchetti, L. Zefiro et al., Am. Miner. **83**, 185 (1998)
9. D. Atencio, J.M.V. Coutinho, S. Graeser et al., Am. Miner. **89**, 1087 (2004)
10. E. Staritzky, A.L. Truitt, in *The actinide elements* ed. by G.T. Seaborg, J.J. Katz (McGraw-Hill, New York, 1954)
11. S.I. Wol'fkovich, E.E. Zuser, P.E. Remen, Proc. Acad. Sci. USSR, Chem. (in Russian) 571 (1946)
12. G. Beadie, M. Brindza, R.A. Flynn et al., Appl. Opt. **54**, F139 (2015)
13. T.A.F. König, P.A. Ledin, J. Kerszulis et al., ACS Nano **8**, 6182 (2014)
14. N. Sultanova, S. Kasarova, I. Nikolov, Acta Phys. Pol. A **116**, 585 (2009)
15. J. Rheims, J. Köser, T. Wriedt, Meas. Sci. Technol. **8**, 601 (1997)
16. D. Węgłowska, P. Kula, J. Herman, RSC Adv. **6**, 403 (2016)
17. N. Suzuki, Y. Tomita, Opt. Express **14**, 12712 (2006)
18. J.-Q. Liu, M. Ueda, J. Mater. Chem. **19**, 8907 (2009)
19. H.-J. Yen, G.-S. Liou, J. Mater. Chem. **20**, 4080 (2010)
20. E.K. Macdonald, M.P. Shaver, Polym. Int. **64**, 6 (2015)
21. A. Javadi, Z. Najjar, S. Bahadori et al., RSC Adv. **5**, 91670 (2015)
22. D. Węgłowska, P. Kula, J. Herman, RSC Adv. **6**, 403 (2016)

Conclusion

Historically, refractometry (and crystal optics more generally) was the first physical method of investigating crystal and molecular structure, and a standard tool for characterising a new substance; hence, acquaintance with it was mandatory for a general chemist. In the last 50 years, explosive growth and automatisation of other physico-chemical methods have largely driven crystal optics out of chemical laboratories, while measuring the refractive indices of *solids* (as distinct from liquids) became almost a lost art. However, refractometry still has an important role to play in a number of fields. Firstly, it is a very efficient method to assess quickly the phase homogeneity of a product and identify different phases, new compounds, and/or polymorphic modifications. RI is a crucial 'fingerprint' of a substance. This is especially important in mineralogy and geology for identifying minerals in field conditions, where the immersion method is indispensable as it requires only a polarising microscope and a set of immersion liquids—eminently portable equipment. Another, prospective field of application is pharmaceutical research, where a development of a new drug now involves a thorough screening for possible polymorphs and phase transitions as an indispensable stage.

On a more theoretical level, refractometry allows to assess the structural motif of a substance; there being an interconnection between the anisotropy of the RI and anisotropy of the atomic structure, including the coordination polyhedra of atoms. Molar refractions, calculated in the form of Lorenz–Lorentz function, are sensitive to the type of chemical bonding. Thus, the ratio of the molar refraction to the molar volume is a measure of bond metallicity (as $n \rightarrow \infty$ for metals) and can be utilised, e.g. to spot a dielectric-to-metal transition under pressure [1]. Applications of molar refraction to solving diverse structural problems are detailed in the book *Refractometry and chemical structure* [2].

Another field of applications concerns the vast area of structural chemistry lying between the individual molecule and the well-ordered crystal. It is noteworthy that standard X-ray diffraction (XRD) technique, today by far the main source of structural information, gives the structural picture averaged over the whole crystal and the whole duration of the experiment. Although it is possible to study local structure (by means of diffuse X-ray scattering) and real-time structural changes (time-resolved experiments), such techniques are still at an early stage of development and are anything but

© The Author(s) 2016
S.S. Batsanov et al., *Refractive Indices of Solids*,
SpringerBriefs in Applied Sciences and Technology,
DOI 10.1007/978-981-10-0797-2

trivial of cheap. XRD is highly efficient for solids with sufficiently high degree of crystallinity, i.e. long-range order, but much less so for amorphous, glassy or semi-crystalline phases and very fine nanopowders – precisely the areas to which the focus of chemical interest is increasingly shifting nowadays. Here, crystal optical methods, with their sensitivity to local structure, can be very useful.

References

1. S.S. Batsanov, A.S. Batsanov, *Introduction to structural chemistry* (Springer, Dordrecht, 2012)
2. S.S. Batsanov, in *Refractometry and chemical structure* (Van Nostrand, Princeton, 1966; *Structural refractometry*, in Russian), 2nd edn. (Vysshaya Shkola, Moscow, 1976)

Printed in the United States
By Bookmasters